Heat and Cold Storage 2

SCIENCES

Energy, Field Directors – Alain Dollet and Pascal Brault

Heat and Cold Storage, Subject Head – Philippe Marty

Heat and Cold Storage 2

Thermochemical Storage

Coordinated by

Nolwenn Le Pierrès

Lingai Luo

iSTE

WILEY

First published 2024 in Great Britain and the United States by ISTE Ltd and John Wiley & Sons, Inc.

ISTE Ltd
27-37 St George's Road
London SW19 4EU
UK

www.iste.co.uk

John Wiley & Sons, Inc.
111 River Street
Hoboken, NJ 07030
USA

www.wiley.com

Library of Congress Control Number: 2024933430

British Library Cataloguing-in-Publication Data
A CIP record for this book is available from the British Library
ISBN 978-1-78945-134-4

ERC code:
PE8 Products and Processes Engineering
PE8_6 Energy processes engineering

Contents

Foreword

Philippe MARTY
LEGI, CNRS, Grenoble INP, Université Grenoble Alpes, France

There has always been an interest in storing heat: prehistoric people heating a stone using fire and then plunging it into a hollow piece of wood filled with water in order to cook their food can be thought of as conducting a first heat storage experiment. As for the Persians, they stored ice in the desert in dome-like buildings with thick clay walls.

Nowadays, heat storage is an important topic of debate on the development of renewable energies. It is generally accepted that massive deployment of these energies should be accompanied by significant storage facilities in order to cope with the irregularity of wind or insolation. Many industrial processes release waste heat. According to a study conducted by ADEME (French Agency for Ecological Transition), nearly 100 TWh of waste heat are released each year, which amounts to approximately one-third of the needs in this sector. In buildings, the currently developed heat storage methods rely essentially on storing water in balloons.

As for the national power grid, the massive solution that is most commonly implemented involves getting water at altitude and then passing it through turbines. Pumped hydroelectric energy storage (PHES) therefore stores electric energy in a gravitational form. This book will not discuss this technology. The same is true for electrochemical storage in batteries or

Heat and Cold Storage 2,
coordinated by Nolwenn LE PIERRÈS and Lingai LUO.

hydrogen vector-based techniques that are the object of other articles in the SCIENCES encyclopedia.

The two volumes presented here have been written by renowned European experts in the field of heat storage and associated materials. They are exclusively dedicated to heat and cold storage and exhaustively discuss all the possible physical and technical solutions. Sensible heat storage methods, in which the temperature of a body is modified to store or release energy, will be presented. The technique involving phase change will also be presented in detail. Thermochemical or sorption heat storage, which is highly promising due to its larger density, will be dealt with in a separate volume.

Volume 1, *Heat and Cold Storage 1: Sensible and Latent Heat Storage*, focuses exclusively on these two types of energy transfers. The state of the art on the standard applications of storage for the building industry is presented, and long-term storage is given particular attention. Indeed, in a country with no steady insolation, the energy autonomy of a building depends on long-term storage possibility, which is also known as inter-seasonal storage. Underground storage offers in this respect possibilities that will be discussed.

The storage of electricity in the form of heat is studied in the presentation of new high-temperature processes. This promising technology is considered for very high power installations and can be seen as an alternative to PHES or to the use of hydrogen vector as a solution for the temporary storage of electricity.

Phase change materials (PCMs) are an attractive alternative when the temperature differences between the storage material and the external environment are too small. It is important to recall that the principle is using the phase change, most often solid–liquid, to accumulate or release the heat transferred in latent form. Due to the high value of the enthalpy of phase change of certain materials, compared to their heat capacity, storage density can be increased. A detailed presentation of the available materials is proposed. The wide range of applications of PCM for the storage of both heat and cold is rapidly extending. For example, they can be found in ice slurries in the form of encapsulated materials. The stakes in the micro- or nano-encapsulation process, and also in the impregnation of porous media, are huge, as they involve difficult techniques that are presented in detail in Volume 1. Besides a high mechanical and chemical stability, the thermal conductivity of the synthesized materials must be high enough to allow for reasonable energy transfer times. This requirement leads to the incorporation of highly conductive materials in the PCM in order to enhance their performances. Chapter 8 is dedicated to the applications of these materials in the industry and construction fields, including the particular case of storage integrated in solar power plants.

Cold storage is essential in a considerable number of applications, particularly in the food industry. Among the high-performance solutions, storage on hydrates is a future possibility that will be presented in Chapter 7.

The particularly promising case of concentrated solar power plants is also discussed in Chapter 8. Though current power plants essentially use molten salts, therefore relying on the principle of sensible storage, future developments of latent and thermochemical systems are presented.

Volume 2 is entitled *Heat and Cold Storage 2: Thermochemical Storage*. Similar to the electrochemical storage being able to sustainably store electrical power, thermochemical or sorption storage is able to store heat or cold over an a priori infinite duration. Unlike sensible or latent storage, this characteristic explains why energy engineers are very much in favor of this technique, which moreover presents a very high energy density that may reach several hundred kWh/m^3. This storage method in the form of chemical potential of the energy includes two distinct classes depending on whether the phenomenon in question is physical (van der Walls forces) or chemical (covalent bonds).

This volume introduces first the principles of sorption in a highly didactic manner and defines the criteria for the selection of materials to be implemented. This is followed by a detailed description of how an absorption cycle operates, which will indicate the future stakes of this method. The challenges that the researchers and manufacturers in the field are currently faced with relate to developing new designs of exchanger reactors, devising new cycles, and implementing new materials.

Particular attention is given to adsorption processes: after a review of the principles of this phenomenon and a thorough description of solid–gas adsorption cycles and systems, a large number of examples of prototype installations are presented.

These two volumes are addressed to several communities: researchers will find here a summary of the progress made by various universities in the field of storage, while academics will find a didactic presentation of the processes implemented. Engineers in research centers will find out the current technological stakes. There is room for improvement in many sectors: materials and associated manufacturing and characterization techniques, equipment, heat exchangers, new cycles, etc.

1

Materials for Thermochemical and Sorption Heat Storage

Kokouvi Edem N'TSOUKPOE
2iE, Ouagadougou, Burkina Faso

1.1. Introduction

It is easy to understand what a material for storage in the form of chemical potential is: we can find examples in our everyday experience. Just take a look at an application that dates back to the Stone Age: the preparation of whitewash using lime obtained from quicklime. What happens when you mix quicklime with water? We can observe that *heat is released*, which becomes all the more intense when a greater quantity of quicklime is mixed with water. The water may start to boil and give off lime, which has a very base pH level and is in fact corrosive. This is why we are often told to pour the lime powder into the water gradually, and not vice versa. After the reaction of quicklime with water is complete, a paste of slaked lime is obtained. What happens next if we dry this paste by bringing it up to a high temperature? The paste *releases water*, and we end up with the same quicklime we started with. It is therefore possible to wet it again and obtain a release of heat. In summary, when you have quicklime in a bag at room temperature, you can say that you have heat at hand, since at any time when you want to obtain heat, you can just mix it with water. It does not matter how long the lime is stored for; the main issue is to protect it from moisture and air. In this way, the drying process, during which you apply heat to allow the slaked lime to release water and thus to *regenerate* the potential for *water absorption* of the lime, can be considered as a process of

Heat and Cold Storage 2,
coordinated by Nolwenn LE PIERRÈS and Lingai LUO.

recharging the lime with heat. This means that lime is a material that allows heat to be stored.

Quicklime is a powder that is essentially made up of calcium oxide (CaO), while slaked lime is made up of calcium hydroxide ($Ca(OH)_2$). The chemical reaction that occurs during the hydration operation is the following:

$$CaO\ [\text{solid}] + H_2O\ [\text{liquid}] \rightarrow Ca(OH)_2\ [\text{solid}] + \text{heat} \quad [1.1]$$

If the amount of water used is not too great – an amount no greater than what is necessary for it to be *hydrated* (1 L of water for every ~3.1 kg of pure quicklime), the slaked lime that is obtained is also in the form of a powder. This result is generally obtained by spraying the quicklime with water. The amount of heat released is $\Delta H = -65\ kJ \cdot mol^{-1}$ of CaO.

Because of its "thirst for water", lime is sometimes spread onto soils that are too humid in order to reduce their water content, and thus facilitate working on them: it is a *hydrophilic* material.

Another hydrophilic material that can be found frequently in our daily lives is silica gel ($SiO_2 \cdot nH_2O$). It is made up of small translucent pellets that can be found in small packets, which are placed in shoe boxes, electronic equipment, or in bottles containing medical tablets. By absorbing the surrounding moisture, these pellets protect the products from damages that might be caused by the presence or introduction of moisture during the transporting or storage of the product. The water attraction capacity or water sorption capacity of the silica gel is due to its large specific surface area; in other words, the area of the real surface relative to its mass. This area is equal to 600–800 $m^2 \cdot g^{-1}$ (Sun and Meunier 2003), meaning that every 10 g of silica gel, the amount contained in a small spoon, has a surface area equivalent to that of a soccer field (which totals roughly 7,000 m^2 for international competitions)! How is this possible? Apart from the size, shape and volume, which have an obvious impact on the *specific surface area*, irregularities or roughness on the surface can greatly influence the actual surface area offered by a particle, and therefore its specific surface area. To demonstrate this, let us consider the beads from the same material, as shown in Figure 1.1. The sizes of the irregularities have been exaggerated for ease of viewing, since at the macroscopic scale, observers perceive them as identical spherical balls (shown with a dotted outline). Thus, the bead presented in Figure 1.1(b) has an irregular contour, which is longer than the bead shown in Figure 1.1(a), and the bead (b) contains a larger external surface area than the ball (a). This is considered to be external surface roughness. In the case of the bead (c), there are also "internal" surfaces connected to the external surface by "passages" which are known as *πόρος* (*poros*) in Greek, a name which also evokes words such as "portal", in the sense of a passageway, or "to carry", referring to transportation. Because of this, we use the word *pores* to refer to cavities, canals, or ducts that are deeper than they are wide (Rouquerol et al. 1994), according to the definition established by the International Union of Pure and Applied Chemistry (IUPAC). Bead (d) contains a certain number of closed pores, which are totally isolated from the others in such a way that no external flow of fluids can access them, and no external gas adsorption can occur there. These closed pores do not contribute to the specific surface area, which focuses on accessible surfaces (Rouquerol et al. 1994).

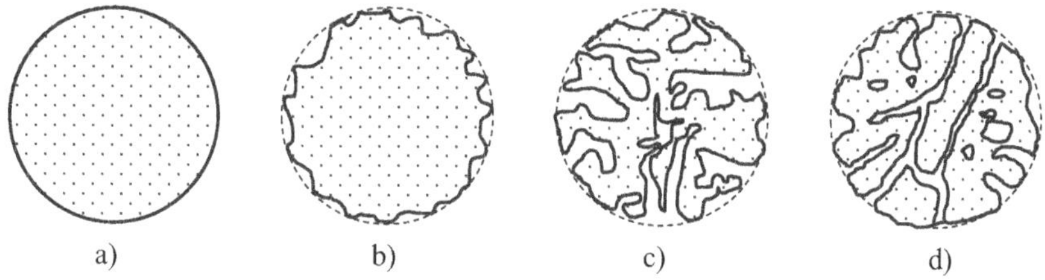

Figure 1.1. *Origin of the specific surface area for a grain of a fixed size of a given material*

One last everyday experience – it depends a lot on where you are! – is the ignition of charcoal. In some large African cities, especially in the West African coastal countries, cooking is primarily done using charcoal as fuel. For ignition, the charcoal is lightly soaked with lamp oil, and it needs to be quickly ignited with a match – or another igniter – otherwise, the flame does not catch and the charcoal quickly seems dry. Indeed, the oil used as lighter fluid is quickly "absorbed" by the coal and disappears in a short time. It is then necessary to soak it again when the fuel is on the outer surface, or not too far down in the pores. This experiment shows that while the charcoal is not specially prepared in such a way as to have a high sorption capacity, it is sufficiently porous and absorbs oil quickly. On the other hand, if any water is accidentally poured onto the coal, the moisture remains apparent (the coal appears wet) and it must then be dried until the coal is fully dry. Coal has a high affinity for lamp oil but not for water.

This chapter is devoted to materials for thermochemical and sorption storage, and begins with the presentation of the key concepts and terminology used in the study of storage materials. It continues with a presentation and a critical analysis of the main criteria for the selection of materials before a presentation on the thermodynamic equilibria of these materials. The next section addresses the materials currently used in the processes developed up to this time, with a particular emphasis on those that are used most frequently. The chapter ends with an introduction to the problem of mass and heat transfers in solid–gas systems and an overview of the material classification techniques for thermochemical and sorption storage.

1.2. Definitions and key concepts

Chemical energy storage uses invertible chemical reactions[1] to store heat. Depending on the form of the energy supplied to the reaction, we will use the terms thermochemical storage (thermal energy), electrochemical storage (electrical energy) or photochemical storage/photosynthesis (electromagnetic radiation) (Figure 1.2).

Sorption is the retention of gas (the sorbate) by a solid or a liquid (the sorbent). It can involve either absorption or adsorption. The latter, a surface phenomenon, is generically defined as the retention of gas or liquid by a solid or a liquid by surface adhesion. More typically, it is the retention of gas at the surface of a solid or a porous material. We use the terms chemical adsorption (chemisorption) when the phenomenon involves covalent bonds, and physical adsorption (physisorption) when it instead involves Van der Waals forces. Absorption refers to a phenomenon in which a substance (a liquid or gas) penetrates into another (a solid or liquid). In heat storage applications, absorption generally corresponds to the absorption of a vapor by a liquid. Chemical sorption involves more energy than physical sorption. It is characterized by a variance equal to 1. Thus, a single state variable (such as temperature or pressure) is all that is needed to describe a chemisorption

1 A reaction is said to be invertible if it can occur in two opposite directions ($A \rightarrow B$ or $B \rightarrow A$) by modifying/reversing the operating conditions, particularly the pressure or the temperature in the cases that interest us here, without going through the same intermediate equilibrium states. This term is used here to avoid confusion with the term "reversible", which is a notion specific to thermodynamics.

equilibrium, while two are required (such as temperature, pressure, or concentration of the sorbate in the sorbent) to describe a physical sorption equilibrium. Section 1.4 will be devoted to the thermodynamic equilibria that are relevant to thermochemical heat storage materials.

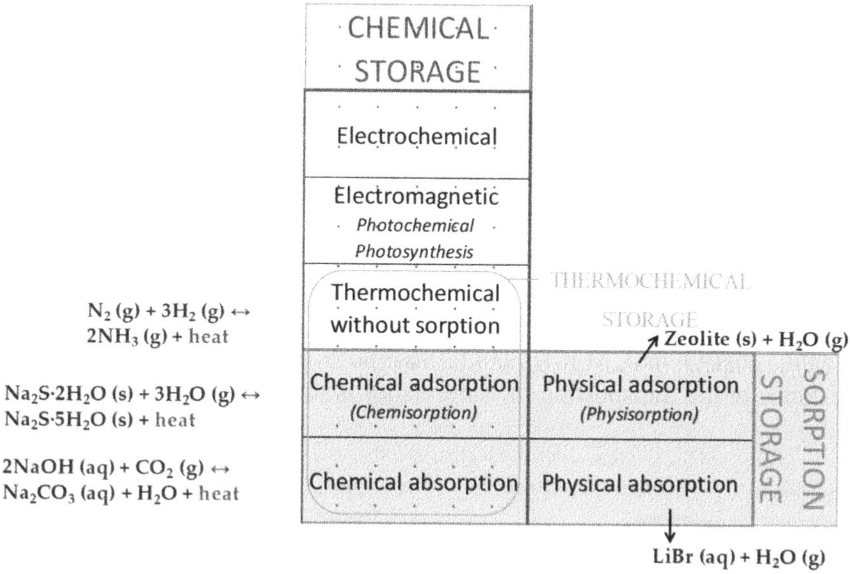

Figure 1.2. *Classification of chemical storage and sorption storage using illustrations: s = solid; aq = aqueous solution; g = gas. Some phenomena, such as storage through photosynthesis, are not reversible*

Several adsorbents are obtained through the agglomeration of microporous crystals to obtain macroporous pellets or granules (Ruthven 1984; Sun and Meunier 1987; Thomas and Crittenden 1998). In this arrangement, the micropores are inside crystals or microparticles, while macropores exist between the crystals, which is to say that they constitute the inter-crystalline space. In the case of the physisorption, depending on whether the physisorbent has one level of porosity (macropores, for example silica gel and active coal) or two levels of porosity (macropores and micropores, for example zeolite), the sorption mechanism is not the same. In the case of a double porosity physisorbent or bidisperse physisorbent, macropores are useful for transporting the sorbate towards the micropores, which provide adsorption capabilities (Sun and Meunier 1987; Silva and

Rodrigues 1999). Figure 1.3 illustrates the general principle of the adsorption mechanism for a physisorbent with double porosity. The adsorbable material, which is to say the potential adsorbate that is not yet adsorbed, diffuses on the surface of the grain with a possible surface resistance. Then it diffuses into the macropores between the particles. The adsorbable reaches the surface of the microparticles and diffuses into the micropores before the intrinsic adsorption on the surface occurs.

For thermochemical reactions such as those involving salt hydrates, which are by far the most commonly used materials in thermochemical energy storage applications for residential applications, an additional phenomenon in in how they are transported must be taken into account. The water is transported through the crystal lattice, forming or breaking bonds. For the dehydration of various salts, a rearrangement of the crystal lattice has been observed, as illustrated in Figure 1.4 for the reaction of the change of $MgCl_2 \cdot 6H_2O$ into $MgCl_2 \cdot 2H_2O$ (Sugimoto et al. 2007). The resulting resistance in the transporting of water vapor is limiting when the crystals are thick.

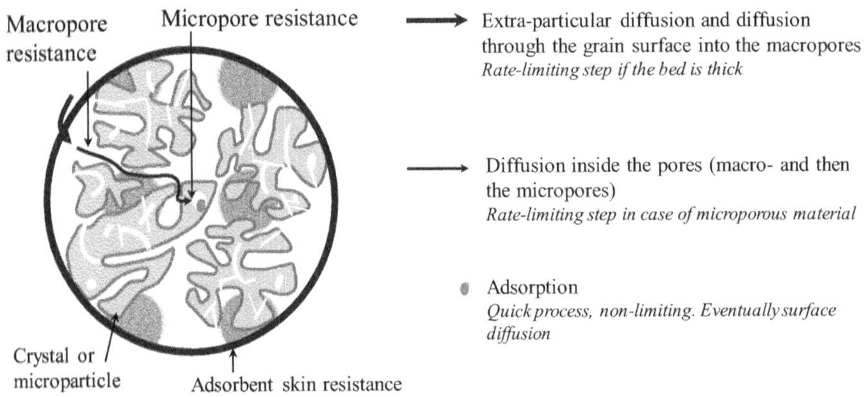

Figure 1.3. *Adsorption in a physisorbent with double porosity. For a color version of this figure, see www.iste.co.uk/lepierres/heat2.zip*

For the sake of simplicity, the expressions "thermochemical storage" and "sorption storage" will be used here interchangeably unless explicitly stated otherwise.

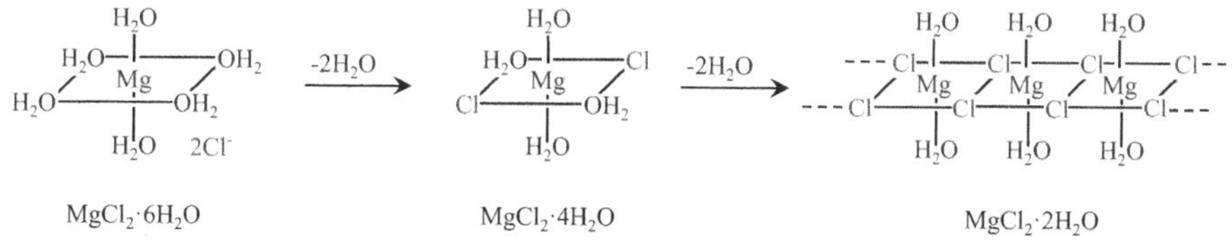

Figure 1.4. *Structural rearrangement during the dehydration of the $MgCl_2 \cdot 6H_2O$ to $MgCl_2 \cdot 2H_2O$ (Sugimoto et al. 2007)*

Therefore, sorption storage or thermochemical heat storage is based on invertible physicochemical phenomena:

$$AB + \text{heat} \leftrightarrow A + B \qquad [1.2]$$

Heat is supplied to the compound AB (a sorption couple or solid–gas reactive pair), which dissociates while absorbing heat to produce two compounds, A and B (equation [1.2] in the forward direction), which can be stored separately: this is an endothermic phenomenon. When these two products are brought into contact, the reverse phenomenon can occur, causing the reformation of the compound AB, releasing heat. During the charging period, heat is supplied to the compound AB in the reactor. Compound B is released in the form of a gas, but in general it is condensed to reduce the storage volume in the case of closed systems. In open systems, compound B is "stored" in the atmosphere, and its volume is not taken into account in the evaluation of the storage density. Since the energy is stored in the form of chemical potential, there are no thermal losses during the storage phase, that is, after the end of the charging and before the start of the discharging; any possible thermal losses are only related to the sensible heat which may have been accumulated in the reactants and are therefore theoretically low, even if the tanks are at room temperature. At the time of the heat demand, the vapor compound B is transferred to the reactor and the compound AB is formed with the release of heat.

In practice, in the thermochemical heat storage processes that have been developed up to now, in almost all cases the sorbate that is used is water because of its many useful qualities, particularly its high availability, its low cost and especially its lack of toxicity, as most of the processes have been developed for "low temperature" applications in households. Thus, the vast majority of research on thermochemical heat storage was focused on long-term applications for housing. Even in high temperature applications, water is the predominant sorbate, although other fluids are also used. For example, the reaction between quicklime and the sorbate CO_2 is considered for thermochemical heat storage in photovoltaic power plants (Fernández et al. 2019) with desorption temperatures reaching up to 950°C.

1.3. Material selection criteria and review of important characteristics for a thermochemical heat storage material

1.3.1. *Selection and overview of key material properties by application*

Regardless of the energy storage method, the selection of the storage material is seen as an important step in the design of the storage process, because this choice affects both the technical and economic performance of the system. Suitable material must therefore store heat without loss and be able to immediately release it on demand at the desired temperature level, and also not generate any harmful effects. The technical, economic or ecological criteria that generally apply to any type of energy storage are (N'Tsoukpoe et al. 2021):

– the thermal storage density, which indicates the amount of energy stored per unit of volume or mass. Mass is an important data point for applications in terms of material costs. The volume becomes more important in cases such as when the system must be installed in a reserved area or when space is expensive, especially in urban areas. It also has an impact on the cost of the installations, especially when it comes to maintaining excessive operating pressures. The thermal storage density is highly dependent on the operating temperatures;

– high thermal conductivity for a good heat transfer and high specific power;

– environmental safety, non-toxicity, a low global heating potential and a low ozone depletion potential. This component involves aspects concerning the intended use of the material at the end of the system's lifespan and the social impacts of its operation and availability;

– the non-corrosiveness of the materials with respect to those commonly used in the design of system components (tanks, piping, etc.);

– good thermal and chemical stability and good cyclability under operating conditions (temperature, pressure): the material should not lose its properties, in particular thermophysical/thermochemical properties after successive cycles of charging and discharging;

– a moderate level of the operating pressure range: if possible, avoid excessive pressure conditions, high vacuums or high pressures;

– low costs for the material: this not only applies to obtaining the material, but also to the costs induced over the entire lifecycle of the process. For example, the consumption of pumps or auxiliaries, and therefore the electrical coefficient of performance, depends on the viscosity, the thermal mass capacity and the density when the storage material is a fluid to be circulated, etc.

Depending on the storage method, other specific criteria may also be considered. For the selection of the sorption pair, in addition to the criteria previously stated, other desirable characteristics of sorption pairs include:

– a sorbent with high affinity for the sorbate, which is important in order to obtain a high specific power;

– thermodynamic equilibrium conditions compatible with the operating temperatures targeted by the process: a regeneration temperature (charging) that is as low as possible and a restoration temperature (discharging) that is high enough to satisfy the needs of the intended use. For a closed system (which will be defined in Chapter 3), the specification of these temperatures is equivalent to the definition of the vapor pressures of the sorbate during charging/discharging. On the other hand, in an open system, the relative humidity of the air is generally used, which is equivalent to a partial pressure of water vapor;

– higher sorbate volatility compared to the sorbent in the case of absorption: this eliminates or reduces the need for purification/rectification of the sorbate released by the endothermic phenomenon;

– the melting temperature of the solid in the case of a solid–gas reaction: often, the material quickly loses its reaction capabilities when it undergoes cycles involving melting.

These criteria are often difficult to reconcile and a compromise is usually necessary.

1.3.2. *Important considerations about certain characteristics and selection criteria for thermochemical heat storage materials*

1.3.2.1. *Binding energy*

Usually, the charging temperature T_c is different from the heat release of discharging temperature T_d. In a thermochemical heat storage process, the charging heat Q_c can be divided into three main parts (Hauer 2007b; N'Tsoukpoe et al. 2012): the binding energy Q_{bind} between A and B, the vaporization heat Q_{vap} of the element B and a portion of sensible heat Q_{sen}. The binding energy is lower in the case of physisorption (Van der Waals forces) than in the case of chemical reactions (covalent forces). Typically, it represents between 15% and 30% of the total heat required for the decomposition of salt hydrates, while representing less of it (5%–15%) in the case of physisorption (N'Tsoukpoe 2018; N'Tsoukpoe and Kuznik 2021). In the latter case, the share of the binding energy depends on the change in the concentration defined by the operating temperatures. The heat needed for the vaporization of the element B typically represents between 65% and 85% of the total heat required for the decomposition of salt hydrates (N'Tsoukpoe et al. 2016; N'Tsoukpoe and Kuznik 2021). The sensible heat necessary to raise the temperature of the material in the reactor from its initial temperature (which is usually ambient temperature) to the charging temperature T_c should not be neglected, especially since the process is not continuous. For chemical reactions with salt hydrates, it represents between 3% and 11% of the total heat required for desorption (N'Tsoukpoe 2018). The proportion of this sensible heat is proportional to the temperature difference between the initial temperature of the reactor and the charging temperature. In the case of an absorption process based on the LiBr–H_2O, this value is more significant (40%), not only because of the temperature difference, but particularly due to the low mixing heat or binding heat Q_{bind} that the phenomena of physisorption present.

The conclusion to be reached is that in a thermochemical heat storage process, the heat actually stored in the process typically represents between 15% and 35% of the total heat required for the decomposition for salt hydrates (N'Tsoukpoe et al. 2016; N'Tsoukpoe 2018), and much less (5%–15%) in the case of physisorption for processes described by different authors (Hauer 2002a; Bales et al. 2005; Fischer 2009;

N'Tsoukpoe 2012, 2018; de Boer et al. 2014; Yu et al. 2014). The thermal efficiency of these processes is therefore affected by this.

1.3.2.2. *Volumetric energy storage density*

The volumetric energy storage density is often crucial, especially for domestic applications, and particularly in cases such as when the system must be installed in a reserved area or when space is expensive, as is the case in urban areas. It also has an impact on the cost of installations from the standpoint of civil engineering. The thermal storage density of the material is not a property of the material: it depends on the operating conditions, particularly the temperatures. For example, considering the pair of $CaCl_2{\cdot}xH_2O$, it is possible to obtain a volumetric energy storage density of 464 $kWh{\cdot}m^{-3}$ of salt under vapor pressure of 12 mbar if the desired discharging temperature is 40°C ($CaCl_2{\cdot}H_2O + 3H_2O \rightarrow CaCl_2{\cdot}4H_2O + \Delta h_r$). This density is only 171 $kWh{\cdot}m^{-3}$ of salt if the discharging temperature is about 60°C ($CaCl_2{\cdot}H_2O + H_2O \rightarrow CaCl_2{\cdot}2H_2O + \Delta h_r$). Thus, the energy storage density offered by a salt can change significantly. Indeed, the hydrates involved in the reaction (for a chemical reaction) or the water contents (for physisorption involving water) needed to obtain the temperature level for the application are not the same.

The volumetric energy storage density can be simply defined as the amount of energy (in this case, heat) that can be extracted from the system under consideration, divided by the volume of the system. With regard to the material, several volumetric energy storage densities can be distinguished depending on the volume considered:

– *absolute volumetric energy storage density*: the reference volume is the volume of solid grains, that is to say the volume of the solid matter alone or absolute volume (Figure 1.5), excluding all voids between grains or inside the grains;

– *the apparent volumetric energy density* per unit of volume of solid matter, including the volume of the empty spaces contained in the grains (closed pores);

– *the bulk volumetric energy density*: this takes into account the entire volume occupied by the bed of particles considered, including consideration for the pores and the volumes of the empty spaces between the particles.

This is the reference density when determining the volume to be allocated to the material in a reactor, for example. A bed of spherical particles of equal diameter arranged following a cubic arrangement has a theoretical intergranular porosity (the share of the spaces between the grains) of 48%, while a rhomboidal arrangement leads to a theoretical intergranular porosity of 26%. Porosity is necessary for good material permeability, but having excessively high permeability is unnecessary, given that at a constant particle diameter, a high porosity generally means a reduced amount of material per unit volume (N'Tsoukpoe et al. 2014a).

Each of the energy storage densities shown in Figure 1.5 should in principle be given with respect to the material (AB, A, or B), which occupies the largest volume. For absorption storage, such distinctions are irrelevant, as the volumetric energy storage density is reported for volumes of liquids.

The question of volumetric energy storage density goes beyond the material, because the end user of the system is more interested in the broader assembly of the system as a whole. In this case, the area under consideration expands to the level of the process or system. At the process level, consideration for the volume of the condensed fluid (in the cases studied in this chapter, water) is a minimum requirement for closed processes. We refer to this as *reduced volumetric energy density*. We can also consider only the heat that is actually stored in the form of a chemical potential or binding energy; we will therefore talk about *effective or net energy density by volume*.

The participants in Task 42/24 of the International Energy Agency (Solar Heating and Cooling programme: Compact Thermal Energy Storage: Material Development and System Integration) agreed on the fact that in order to ensure the competitiveness of thermochemical heat storage, the volumetric energy storage density it offers must be four to eight times higher than that of water (Wong et al. 2013), or approximately 360 ± 120 $kWh \cdot m^{-3}$. However, the boundaries of "energy storage density" remain somewhat unclear in the statement, but one can reasonably assume that it refers to a reduced bulk volumetric energy density.

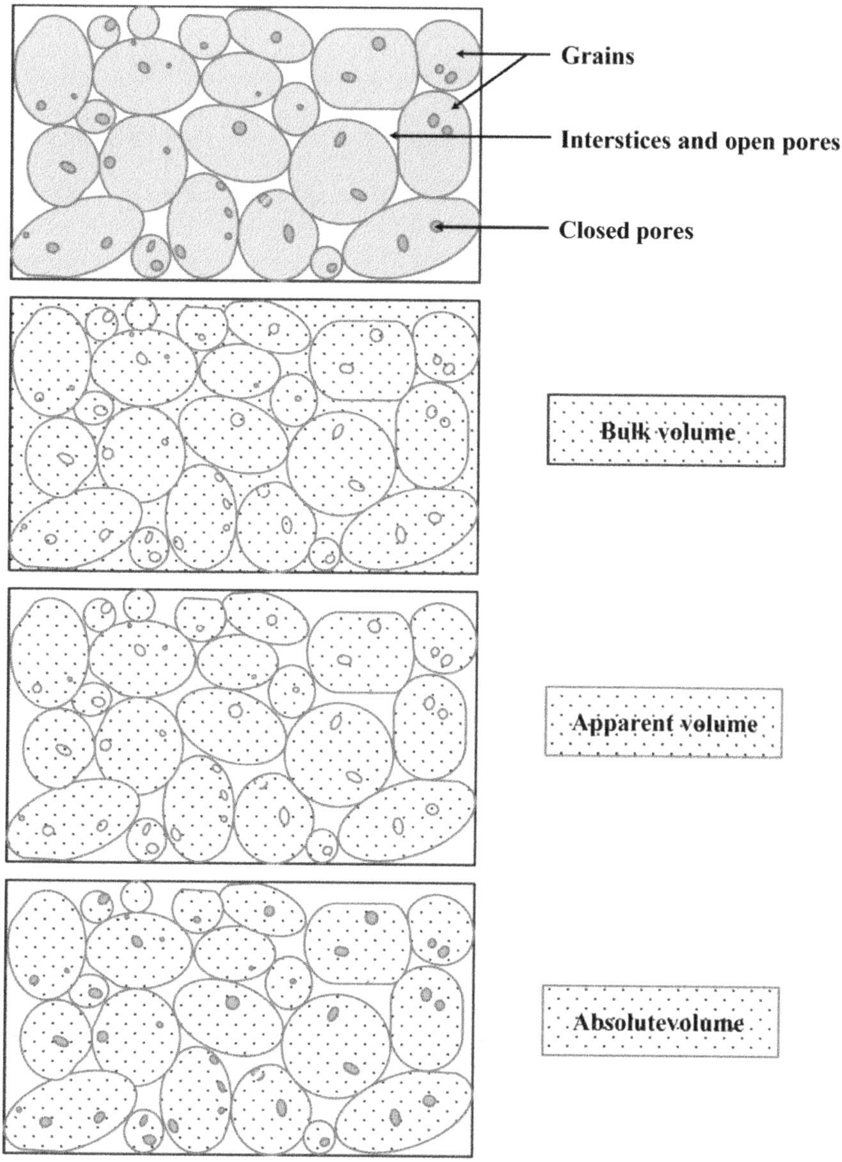

Figure 1.5. *Solid–gas storage: definition of the bulk volume (materials + pores of closed and open pores or gaps between the grains), the apparent volume (grains + closed pores), the true or absolute volume (only the grains outside closed pores) and the volume of the closed pores. For a color version of this figure, see www.iste.co.uk/lepierres/heat2.zip*

1.3.2.3. *The cost of the material*

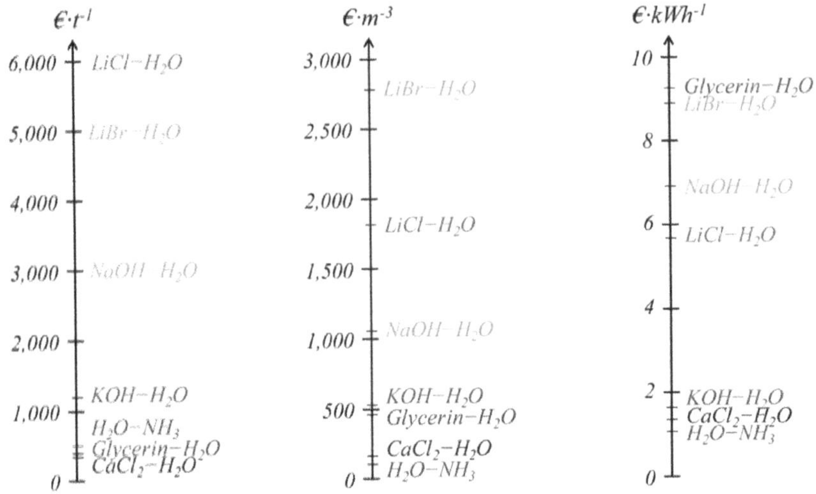

Figure 1.6. *Comparison of the costs for different absorbents in heat storage. For the calculation conditions and price sources, see Table 1.2*

There is currently no definition of a low-cost material. This is mainly due to the fact that the price varies in space and time, as well as the fact that the relevance of the storage will also depend on several local factors, in particular the cost of conventional energy (e.g. the cost of producing heat from fuel oil). To give a clearer idea of this, in the context of Europe at the beginning of the 2010s, the rate of 0.25 €·kWh^{-1} has been deemed as a desirable target rate by various authors (Hauer 2010; Zondag et al. 2010). More recently, within the same context and in various scenarios, this rate was actually 0.60–1.43 €·kWh^{-1}, which was proposed by a group of experts from Task 42 of the International Energy Agency Solar Heating and Cooling Programme (Rathgeber et al. 2015, 2016; Van Helden and Rommel 2015). All such costs are expressed in €·kWh^{-1}, since the common reference to the processes being compared is the amount of stored heat (N'Tsoukpoe et al. 2014b). This makes it possible to take into account the thermal storage density and immediately see the material that will result in a higher investment cost for a given thermal energy storage project. For example, in a study shown in greater detail in section 1.5, the cost of glycerin is 500 €·t^{-1} (Figure 1.6) and allows a storage density of 54 kWh·t^{-1}. Under the same conditions, the LiCl costs 6,000 €·t^{-1} and allows a storage density of 1,053 kWh·t^{-1}. Glycerin is the most expensive material for storage, since for each kWh of heat stored and assuming a single cycle, it is necessary to invest €9.3, while LiCl requires €5.7. These illustrations show that the selection of the material based on the cost of the purchase or acquisition of the material, in terms of the price per unit of mass ($€\cdot kg^{-1}$) or per unit of volume ($€\cdot m^{-3}$), can be misleading. The question of cost should be analyzed over the entire lifecycle of the planned storage, and the important data available should be used for this purpose. A material may very well have a low cost per kWh of stored heat but be highly corrosive and operate under high vacuum, requiring the use of materials resistant to its corrosiveness for the construction of heat exchangers or the reactor, and requiring special components to withstand high vacuum.

1.4. Description of the thermodynamic equilibrium of sorption materials

As indicated previously, the phenomena of chemisorption are monovariant phenomena, while the phenomena of physisorption are divariant. Thus, one state variable (such as temperature or pressure) is all that is needed to describe a chemisorption equilibrium, while two are required (such as temperature, pressure, or concentration of the sorbate in the sorbent) to describe a physical sorption equilibrium. To describe the liquid–vapor or solid–gas equilibria, simplified forms of the Clapeyron equation are used, and the associated diagrams simplify the study of sorption cycles involving these equilibria. In particular, the Clausius–Clapeyron equation (equation [1.3]) is commonly used to describe the liquid–vapor equilibrium or phase change phenomena of a pure substance when far from the triple point, and where the vapor phase can be assumed to be an ideal gas:

$$\frac{dP_{eq}}{dT} = \frac{P_{eq} \cdot \Delta H}{R \cdot T^2} \qquad [1.3]$$

where T is the temperature, P_{eq} is the saturated vapor pressure of the pure substance at the temperature T and ΔH is the specific enthalpy of the phase change at the temperature T.

1.4.1. *The case of chemisorption*

If we assume that the specific enthalpy of the phase change is constant, that is, independent of the temperature, an assumption that can be verified by the small temperature variations that generally occur in sorption thermal storage processes, we may then integrate equation [1.3] to obtain Rankine's formula:

$$\ln\left(\frac{P_{eq}}{P_0}\right) = -\frac{\Delta H_r}{R} \cdot \left(\frac{1}{T_{eq}} - \frac{1}{T_0}\right) = -\frac{\Delta H_{r0}}{RT_{eq}} + \frac{\Delta S_{r0}}{R} \qquad [1.4]$$

Rankine's formula, usually referred to as Clausius–Clapeyron, the formula from which it derives, is generally used to describe the equilibrium conditions of chemisorption phenomena. In equation [1.4], T_{eq} is the equilibrium temperature (of the sorbent, which is to say, solid or liquid), P_{eq} is the equilibrium pressure of the gas (sorbate) above the sorbent, ΔH_{r0} is the enthalpy of the reaction per mole of gas and ΔS_{r0} is the entropy of the reaction per mole of gas. T_0 and P_0 are the temperature and the pressure at the reference point, respectively. ΔH_{r0} and ΔS_{r0} can be determined experimentally, using methods such as thermogravimetric analysis (TGA) in together with differential scanning calorimetry (DSC) (see section 1.7). Alternatively, they can be estimated from the standard enthalpies and entropies of formation of the compounds that can be found in the database of chemical thermodynamic properties of the NBS (currently the NIST) (Wagman et al. 1982). The values ΔH_{r0} and ΔS_{r0} experimentally measured by different authors (Fischer 2009; Dicaire and Tezel 2011; Gaeini et al. 2017) are between 72% and 95% of the value calculated from this database, which is relatively satisfactory for the estimates of the energies involved, in particular for the energy storage density.

The equilibrium points are generally represented in a pressure–temperature diagram, which is known as the *Clausius–Clapeyron plot*. By choosing $-1/T$ for the temperature scale and ln (P) for that of the y-axis, the equilibrium for each reaction is represented by a straight line following equation [1.4], as illustrated in Figure 1.7.

EXAMPLE.– Estimate the equilibrium temperature of the $SrBr_2 \cdot 6H_2O$ under a saturated water vapor pressure at 10°C. First, we identify the curve for the reaction $SrBr_2 \cdot H_2O + 5H_2O \rightarrow SrBr_2 \cdot 6H_2O + \Delta h_r$, and then the saturation curve for pure water in Figure 1.7. Using this second curve, the saturation vapor pressure of water at 10°C could be found, approximately 12 mbar. From this, the equilibrium temperature corresponding to $SrBr_2 \cdot 6H_2O$ is $T_{eq} = 45°C$.

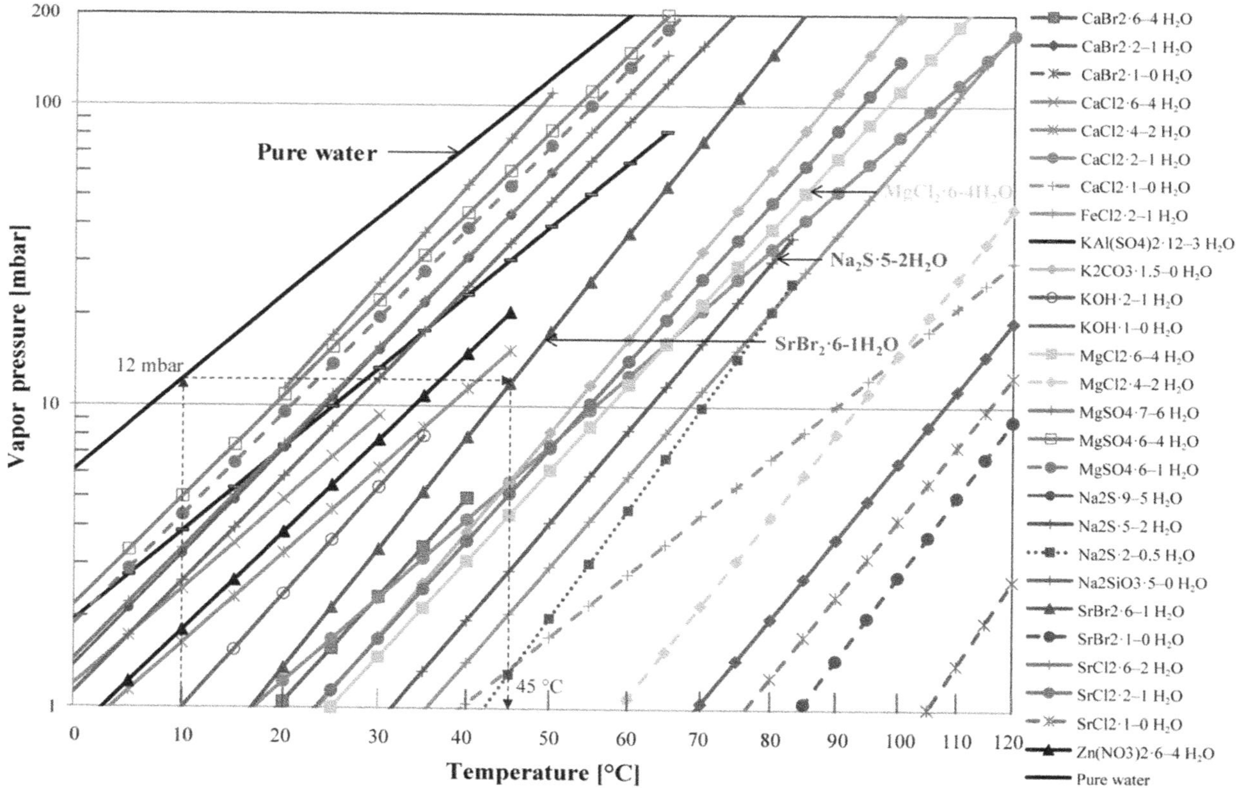

Figure 1.7. *Solid–gas equilibrium curves of salt hydrates in a Clausius–Clapeyron diagram, mostly constructed from the calculated data from the NBS database. For a color version of this figure, see www.iste.co.uk/lepierres/heat2.zip*

1.4.2. *The case of physisorption*

The equilibria of physisorption involve the pressure *P*, the temperature *T* and the composition *X*. This last variable accounts for the amount of sorbate that is sorbed by the sorbent. Since physisorption phenomena are divariant, it is necessary to define two of these three parameters in order to deduce the third; in other words, these equilibria obey laws following the form P = f(T, X).

1.4.2.1. *The case of physical absorption*

In the study of liquid–gas mixtures, the composition is usually taken into account using the mass fraction, which we will also refer to – though some chemical purists will object to this naming! – as concentration or mass content. In a mixture, the mass fraction of a constituent is the mass of this constituent divided by the total mass of all constituents in the mixture, and is denoted as w, according to the recommendation of the IUPAC (International Union of Pure and Applied Chemistry 2014). For example, when we dissolve 30 g of anhydrous lithium bromide (anhydrous LiBr) in 70 g of water, we obtain a solution whose mass fraction in LiBr is $w_{LiBr} = 0.3$. By multiplying this number by 100, we obtain the mass percentage, which is denoted w%, or in this case: $w\%_{LiBr} = 30\%$. Other formulations of properties (such as Patek and Klomfar (2006)) use the molar fraction instead, which is denoted x, that is to say the ratio of the number of moles of the constituent considered relative to the total number of moles. It should be noted that not all authors adopt the same notation, and that it is therefore necessary to carefully check the definitions of the variables used. For example, in this chapter, the molar fraction is denoted X. In the study of absorption processes involving binary mixtures (Chapter 2), the concentration of a solution is usually given relative to the most expensive constituent. Thus, for a LiBr–H_2O mixture (aqueous solution of lithium bromide) where the water is the sorbate, the concentration will instead be given by referring to the mass fraction of the salt (lithium bromide). On the other hand, for the absorption pair H_2O–NH_3 (aqueous ammonia solution) where the ammonia is the sorbate, the concentration will instead be given by referring to the mass fraction of ammonia.

The Clausius–Clapeyron equation also applies to absorption equilibria by treating a mixture of a given concentration as a pure substance. Indeed, the enthalpy and the entropy of the sorption depend on the concentration of the solution. The representation of these constant composition lines in the (ln(P) *versus* −1/T) plane gives us the *Oldham diagram* of the mixture. In this diagram, the constant composition lines are therefore very close to straight lines, which makes them easier to interpolate. Figure 1.8 shows the Oldham diagram of the NaOH–H_2O pair (an aqueous solution of sodium hydroxide or caustic soda).

Another way of representing the binary mixture equilibria is to use the *Dühring plot*. It translates Dühring's rule, which implies that for a given pressure, the equilibrium temperature of the solution of the binary mixture is a linear function of the saturation temperature of the pure solvent at the same pressure. It follows that in a diagram in which the temperature of the solution is on the *x*-axis and the saturation temperature of the solvent is on the *y*-axis, the constant composition lines are parallel straight lines.

Compared to the Oldham diagram, it has the advantage of using linear coordinates (the temperatures) while maintaining the linearity of the constant composition lines. A secondary axis can be added to give information on the saturation pressure corresponding to the sorbate. It can be observed that Dühring's rule also derives from the Clausius–Clapeyron relation. For the aqueous solution of LiBr–H_2O, for example, the slope of the straight lines is approximately 1.1, which can simplify some thermodynamic analyses of the processes (Anies 2011). Figure 1.9 shows the Dühring's plot of the absorption pair LiBr–H_2O.

1.4.2.2. *The case of physical adsorption*

In the case of physical adsorption, it is also possible to represent the pressure equilibria with an Oldham or Dühring's plot (Figure 1.10). As we have already noticed, this type of visualization facilitates interpolations and is well suited to represent a sorption process, and thus a thermal storage cycle based on sorption.

But here, the concentration is generally the content of the adsorbate, which is to say the amount of adsorbate absorbed per unit of dry mass of adsorbent. For example, when 30 g of water are taken in by 100 g of dry silica gel, the concentration or the water content is X = 0.3 kg of water/kg of silica gel.

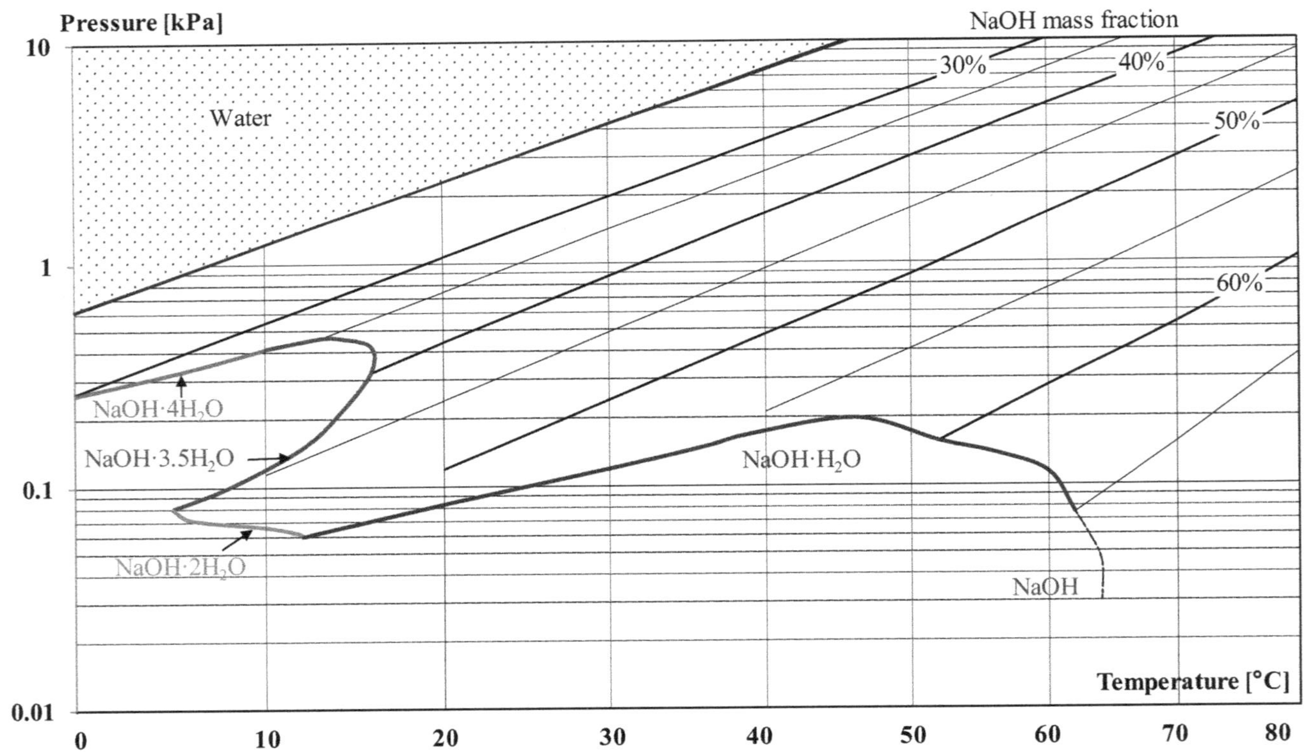

Figure 1.8. *Oldham diagram of the NaOH–H_2O pair. For a color version of this figure, see www.iste.co.uk/lepierres/heat2.zip*

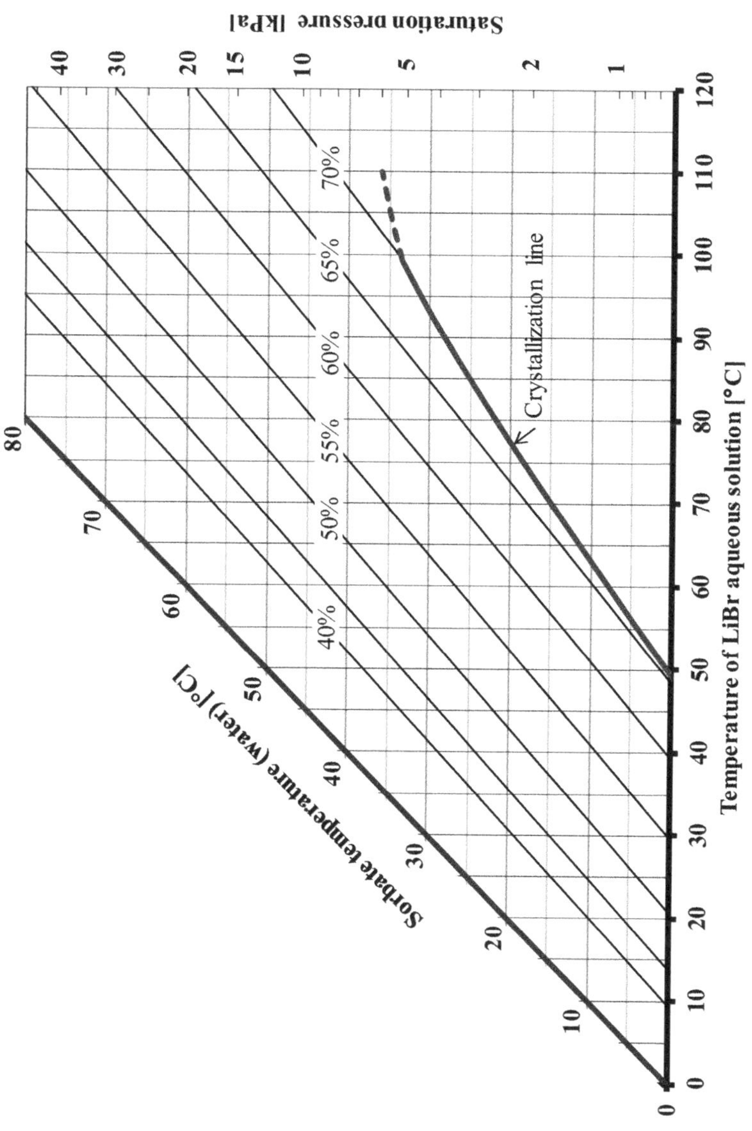

Figure 1.9. *Dühring diagram of the LiBr–H_2O pair*

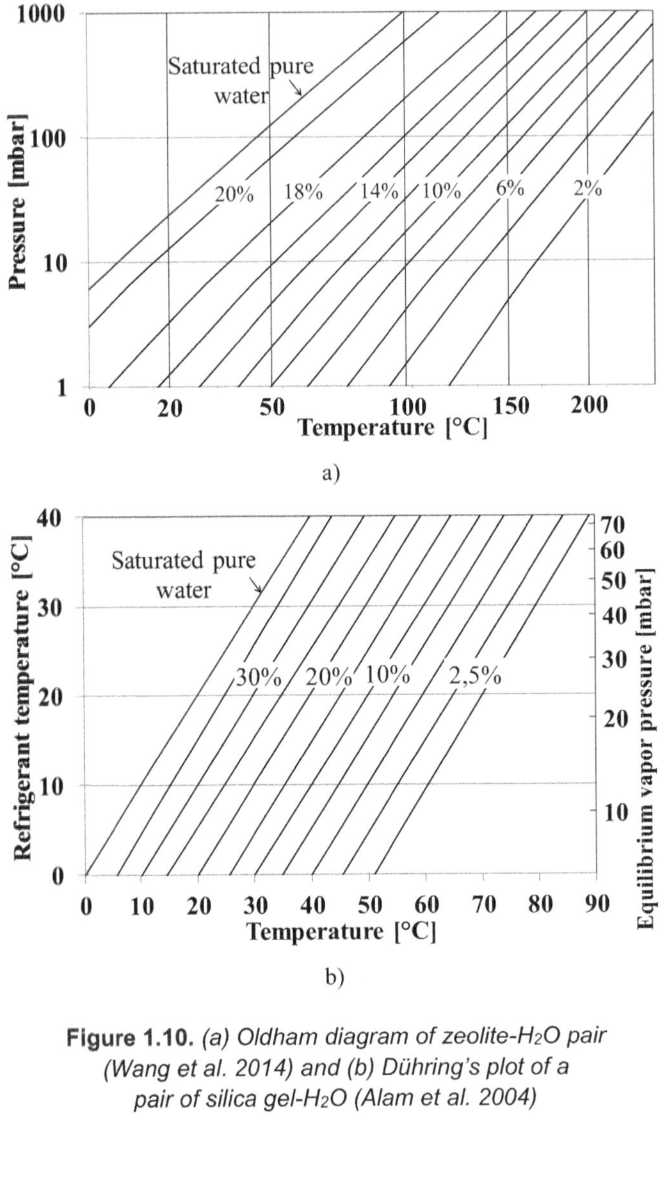

Figure 1.10. *(a) Oldham diagram of zeolite-H_2O pair (Wang et al. 2014) and (b) Dühring's plot of a pair of silica gel-H_2O (Alam et al. 2004)*

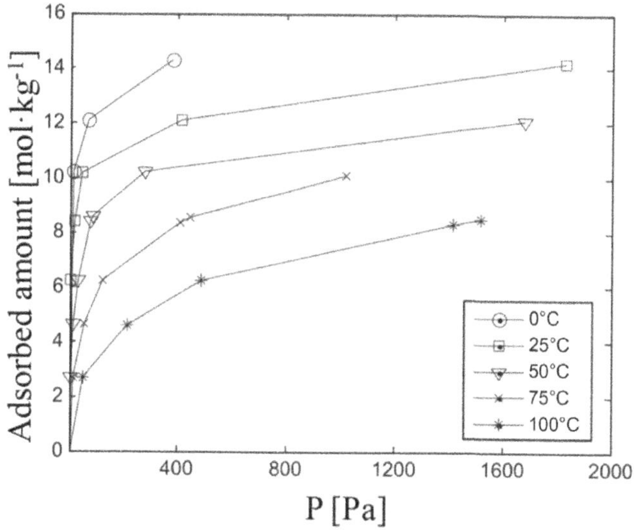

a)

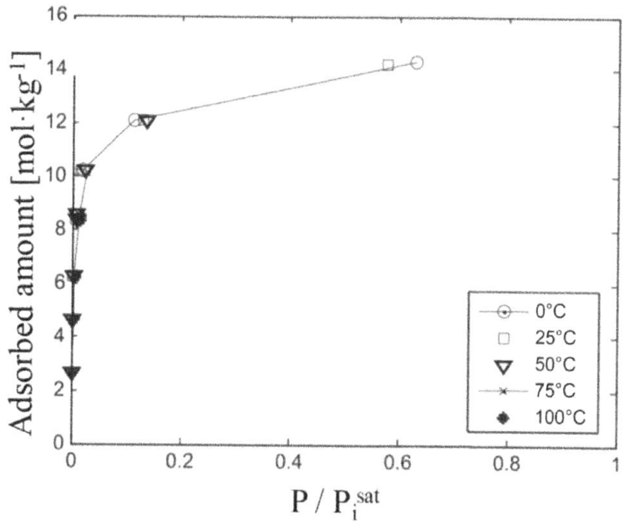

b)

Figure 1.11. *(a) Sorption isotherms of zeolite-H_2O pair; (b) representation of the same points in a diagram introducing the saturation pressure of water (Leppäjärvi et al. 2012)*

Another common way of representing adsorption equilibria, although less convenient for illustrating a sorption cycle or a storage process, is the use of sorption isotherms. Their frequency is linked to the fact that physisorbents are widely used in gas separation processes with adsorption columns operating under isothermal conditions. In this graphical representation, pressure and concentration constitute the axes of the diagram, and the variations at constant temperature – the isotherms, therefore – are plotted (Figure 1.11(a)). There are six main types of adsorption isotherms (Rouquerol et al. 1999), depending on the shape of the isotherm and therefore on the characteristics of the adsorbent. Mathematical models associated with each type of isotherm allow for proposing correlations to represent experimental data as accurately as possible while maintaining physical significance (Sun and Meunier 2003), with model parameters determined experimentally for each adsorption pair. Intermediate isotherms and compound isotherms exist; in fact, this is most often the case, and different types of isotherms can be used to describe parts of an experimental isotherm.

To simplify the models for the simulation, correlations that do not necessarily follow the basic models can be developed. For example, equation [1.5] proposed to describe the sorption equilibrium of the zeolite 13X–H_2O (Tatsidjodoung et al. 2014):

$$X_{eq} = a_2 \cdot T^2 + a_1 \cdot T + a_0 + 0.0228 \cdot \ln\left(\frac{P}{P_{sat}}\right) \qquad [1.5]$$

with X_{eq} being the content of water at equilibrium, T the temperature of the zeolite, P the pressure of the water vapor and P_{sat} the saturation pressure of the water at the temperature T. $a_2 = 3 \times 10^{-6}$ K^{-2}, $a_1 = -0.0017$ K^{-1} and $a_0 = 0.3206$.

In some cases, the data can be reduced to a single curve when the saturation pressure of the pure adsorbate is involved (Leppäjärvi et al. 2012): the unique curve obtained is then independent of temperature. The illustration is shown in Figure 1.11(b) where all the isotherms presented in Figure 1.11(a) are reduced to a single curve. The trick is to switch from the water vapor pressure to the relative pressure defined as the ratio of the water vapor pressure P to the saturation pressure of water at the temperature of the adsorbent, which is equivalent to relative humidity in the case of humid air (Bales et al. 2005).

1.5. Overview of the main materials studied in the context of thermochemical energy storage

Section 1.5 provides an overview of various materials studied in the context of thermochemical heat storage. The goal is not to present an exhaustive list of studied materials but rather to highlight the most representative ones or those that seem to stand out within the considered category.

1.5.1. *Sorbates*

Theoretically, several fluids could be suitable as sorbates in sorption processes (Le Pierrès 2014): water, ammonia, alcohols, alkanes, hydrocarbons, amines, etc. In practice, only water and ammonia are generally used. In a survey of commercially available sorption refrigeration machines compatible with a solar thermal installation, out of around 60 machines, 90% used water as the sorbent, and the remaining 10% operated with ammonia (Mugnier 2013). More specifically, in the context of thermochemical heat storage, water appears to be the fluid of choice; at least for domestic applications, it is the "king" of the sorbates. Its main advantages are obvious.

Table 1.1 presents a comparison with ammonia in the perspective of a thermochemical heat storage application. Ammonia is generally ruled out mainly due to the risks of inhalation poisoning, as even though the odor of this gas is pungent and putrid, making its leak detectable from as low as 5 ppm, intoxication during sleep or an inability to escape is not to be excluded.

Each year, several cases of deaths attributable to inhalation poisoning by gaseous or liquid ammonia are recorded in France, both in professional and domestic environments (Darchy et al. 1997). Apart from chlorine, ammonia is among the gases most frequently involved in accidental domestic poisonings (Tournoud 2008). Although widely used in industry, ammonia is deemed too dangerous for domestic use, and its storage close to places likely to accommodate a large number of people is regulated (Office Fédéral de l'Environnement 2015).

Water	Ammonia
Inexpensive material (++) Non-flammable (++) Non-toxic (++) Zero potential destruction of ozone layer (++) Operation at low pressure, meaning pressures lower than atmospheric pressure in closed systems: limitation of heat and mass transfers, especially in solid–gas reactors, potential sealing problems (–) Suitable thermophysical properties (+) High heat of vaporization, at about two times that of ammonia (++) Compatibility with a large number of materials used for civil engineering Usable for open systems	Inexpensive materials (+) Flammable fluid (–) Toxicity to humans and the environment (– –) Storage subject to regulations (declaration or authorization) (–) Operation at pressure above atmospheric pressure (±) Suitable thermophysical properties (+) High heat of vaporization, though it is two times lower than that of water (+) Corrosive to copper and copper alloys (–) Impossible to use in the case of open systems

Table 1.1. *Comparison of water and ammonia in the perspective of a thermochemical heat storage application*

Thus, most of the thermochemical heat storage processes are water based. For open processes, as water vapor is naturally present in the atmosphere, this means that once again, water is the preferred fluid. Though water is still the preferred fluid in almost all thermochemical heat storage processes for industrial applications, there are some cases where other fluids are considered. This is the case for CO_2, which has been proposed for thermochemical heat storage in photovoltaic power plants with desorption temperatures of up to 950°C (Fernández et al. 2019).

1.5.2. *Sorption pairs*

1.5.2.1. *Physical absorption*

The absorption pairs considered for sorption heat storage are initially those already used or studied for sorption-based cooling or heating processes (heat pumps). After experiments on these classic sorption pairs, which are relatively few in comparison to the number of studies focused on chemical reactions, there have been some efforts to develop materials specifically tailored for thermal storage by sorption. The first studies on the selection of materials for thermal storage by absorption examined and compared different absorption pairs, such as $LiBr–H_2O$, $NaOH–H_2O$, $CaCl_2–H_2O$, glycerin–H_2O, KOH–H_2O, LiCl–H_2O and H_2O–NH_3 (Liu et al. 2011; N'Tsoukpoe 2012). The main points of comparison between the pairs studied are summarized in Table 1.2. The calculations of the energy storage density are carried out for the same operating conditions (which will be explained in Chapter 2):

– condenser temperature = 30°C;

– minimum heat production temperature in the absorber = 20°C;

– storage tank temperature = evaporator temperature = 10°C.

The detailed approach of the evaluation of the energy storage density is presented by N'Tsoukpoe (2018). The value of 20°C used as the minimum temperature to be provided by absorption is objectively low to meet heating needs. This is explained by the fact that it is the temperature level allowed by considering the $CaCl_2$–H_2O pair. The classical efficiency γ (Table 1.2) is the heat delivered by absorption divided by the heat required for desorption. As previously defined, the reduced volume storage density is the one taking into account the volumes of the tanks containing the solution and water, at their respective maximum capacities during the cycle. The toxicity evaluation is done using the NFPA, which is a standard defining the main risks associated with each product on a scale from 0 (safe) to 4 (very high risks).

An initial comparison is made by considering a crystallization-free storage (Table 1.2). For salt solutions, the maximum concentration in the solution tank is therefore a value close to the solubility limit of the solution at the chosen temperature of the storage tanks, 10°C. For the pairs glycerin–H_2O and H_2O–NH_3, there is no crystallization, regardless of the concentration of the solution. A maximum concentration of 90 w% was decided on, due to the difficulties in reaching a concentration of 100 w%. For the pair H_2O–NH_3, with water as the absorbent, the minimum concentration of NH_3 in the corresponding mixture is therefore 10 w%. The second comparison considers the storage with crystallization in the solution tank, making it possible to increase the energy density storage, which will be discussed in Chapter 2 (section 2.4.1). The prices shown here (Table 1.2) should be considered relative, since they vary over time and space.

Indeed, it is difficult to establish reliable prices for large quantities (several tons to be stored) of products that are not mass-produced. In this case, the price is given for the storage material other than water, considering the same material purity (>99%), on the Chinese market (December 2008).

The LiCl–H_2O and KOH–H_2O pairs have practically the same performance (Table 1.2) in terms of volumetric storage density and efficiency, taking into account the uncertainties on the properties used in the calculations. The disadvantage of the two salts containing lithium is their high cost, likely sustained by the demand for lithium, especially in electrochemical storage applications. The highest viscosity of LiCl raises concerns of lower heat transfer coefficients (Pataskar et al. 2011). The high viscosity also creates a significant load on the solution circulation pumps, which can also be damaged (Pataskar et al. 2011). The mixture of the LiBr and LiCl solutions creates a solution with a higher viscosity than each of them individually but reduces the risk of crystallization for the latter since its solubility range is narrower. Therefore, KOH–H_2O would seem to be the best choice when considering the cost (1.7 €·kWh^{-1}) and the volumetric storage density. Its viscosity is comparable to that of the LiBr solution, but the desorption of this solution requires a higher temperature level.

The LiBr–H_2O pair provides the highest volumetric energy storage density (Figure 1.12) as the heat release temperature increases, with a greater margin for heating (Figure 1.12). For a temperature of 35°C in the absorber, the reduced volumetric storage density is approximately 100 kWh·m^{-3}.

In Figure 1.12, only a few points related to the NaOH–H_2O pair have been represented. Indeed, this solution has a crystallization curve that is in fact very particular (Weber and Dorer 2008; Fumey et al. 2014), and special arrangements are required to practically obtain the point corresponding to 44.5°C (N'Tsoukpoe 2012). The significant issues with severe corrosion and of the particular crystallization curve (Ferchaud 2016) – leading to clogging in pipes and circulators – make this salt less attractive. Figure 1.12 also shows that the storage density becomes zero for a given temperature, generally because the concentrations of diluted and concentrated solutions approach each other, but this is not always the reason. Analyzing closely, for example, the LiBr–H_2O pair, it is observed that beyond this temperature (45.8°C), the "latent" heat (heat of dilution and latent heat provided by the water vapor) is lower than the sensible heat necessary to bring the solution from the temperature of the storage tank to the desired absorption temperature. This implies that the thermal storage by classical absorption based on LiBr–H_2O, or the other pairs of salt solutions presented here, cannot provide heat for DHW applications requiring a temperature of around 60°C (Figure 1.12).

Pair		$CaCl_2$–H_2O	Glycerin–H_2O	KOH–H_2O	LiBr–H_2O	LiCl–H_2O	NaOH–H_2O	H_2O–NH_3
Price (€ per metric ton of storage material)		350	500	1,200	5,000	6,000	3,000	400
NFPA 704	Health	2	1	3	2	2	3	3
	Flammability	0	1	0	0	0	0	1
	Instability/reactivity	1	0	1	0	0	1	0
Storage without crystallization	Concentration after desorption (m%)	39.8	90.0	50.8	58.8	42.8	33.5	90.0
	Desorption temperature (°C)	44.8	53.0	63.0	72.0	63.1	50.0	155.5
	Energy storage density ($kWh \cdot t^{-1}$)	254	54	727	561	1,053	433	366
	Solution energy storage density ($kWh \cdot m^{-3}$)	119	50	313	313	319	154	98
	Reduced energy storage density ($kWh \cdot m^{-3}$)	102	–	217	217	220	126	–

Table 1.2. *Elements of comparison of the various absorption pairs studied*

Pair		$CaCl_2$–H_2O	Glycerin–H_2O	KOH–H_2O	LiBr–H_2O	LiCl–H_2O	NaOH–H_2O	H_2O–NH_3
Storage without crystallization	Cost of storage material (€*kWh^{-1})	1.4	9.3	1.7	8.9	5.7	6.9	1.1
	Pressure during charging (mbar)	42	42	42	42	42	42	1,167
	Pressure during discharging (mbar)	12	12	12	12	12	12	615
	Classic efficiency γ	0.909	0.545	0.830	0.850	0.866	0.750	0.665
Storage with crystallization	Concentration after desorption (m%)	48.5	–	58.8	68.3	51.8	37.8	–
	Desorption temperature (°C)	54.2	–	84.0	91.4	69.4	57.0	–
	Energy storage density (kWh·t^{-1})	562	–	916	725	1,326	662	–
	Solution energy storage density (kWh·m^{-3})	263	–	400	404	401	233	–
	Reduced energy storage density (kWh·m^{-3})	–	–	–	256	255	–	–
	Cost of storage material (€*kWh^{-1})	0.6	–	1.3	6.9	4.5	4.5	–

Table 1.2. *Elements of comparison of the various absorption pairs studied (continued)*

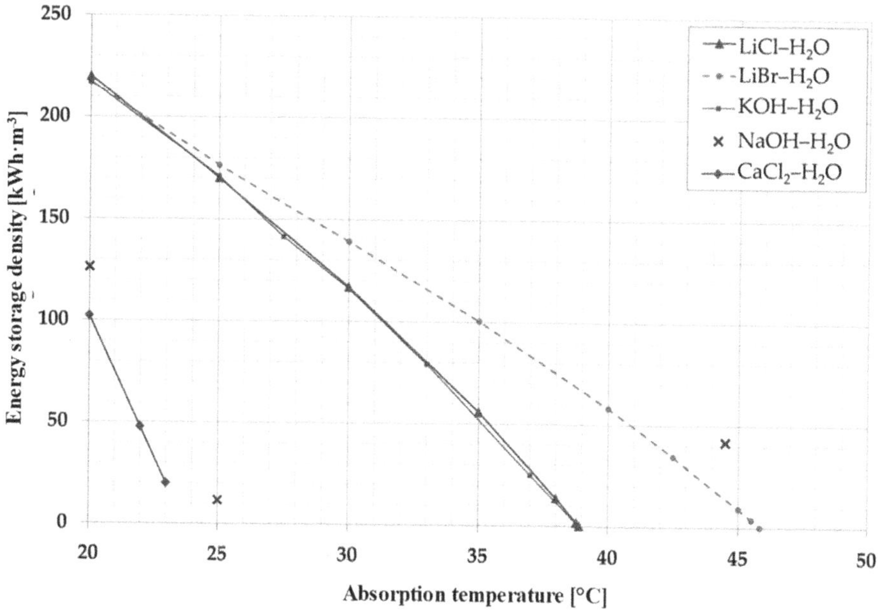

Figure 1.12. *Evolution of the reduced volumetric energy storage density as a function of the absorption temperature for different absorption pairs. Evaporator temperature = storage temperature = 10°C, storage without crystallization (concentration of the solution at the end of charging at the saturation limit). For a color version of this figure, see www.iste.co.uk/lepierres/heat2.zip*

Another limit prevents the use of the aqueous salt solutions currently considered for the production of heat at a temperature level of 60°C, particularly in the case of seasonal storage. This has to do with the saturation curve and the evaporator temperature. Let us consider LiBr–H_2O. If we consider the evaporation pressure corresponding to a temperature of 10°C, the minimum concentration of the solution that would be required would be x = 64.7% at the end of absorption. However, a solution with such a concentration could only be drawn from a storage tank whose temperature is about 40°C (see the Dühring's plot of the pair LiBr–H_2O in Figure 1.10), since this is the saturation concentration of the solution at 40°C. It is difficult to consider maintaining the solution reservoir above such a temperature in ordinary storage, unless the storage of sensible heat is managed with the appropriate insulation. These conclusions also apply in the case of storage using crystallization, which will be discussed in Chapter 2 (section 2.4.1), because the solutions used in the evaluations are already at the saturation point. One way to overcome this limitation could be the use of additives or mixtures that shift the crystallization curve or inhibit crystallization (Adegoke 1993; Lee et al. 2000; Dirksen et al. 2001; Ring et al. 2001; Chan et al. 2013), which could also allow them to increase the energy storage density.

However, the cost of LiBr is high. In addition, the solution that is formed is viscous and corrosive. The conventional method of thermal storage by absorption therefore offers a thermal storage density comparable to that of sensible water storage when the temperature of the intended application exceeds 40°C. New approaches, both at the material and process/system levels, are essential to make this thermal storage process relevant.

To this end, studies have been undertaken to identify or develop inexpensive sorption pairs capable of offering a higher energy storage density (Lefebvre 2015) for a process involving crystallization. From an initial list of 220 salts, a preliminary selection was made based on arbitrary but reasonable values for different criteria initially defined (Figure 1.13).

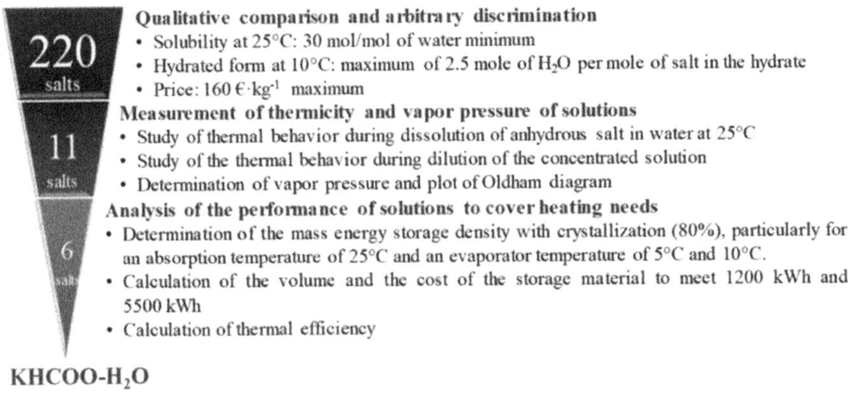

Figure 1.13. *Flowchart of the salt selection process for long-term thermal storage by absorption adopted by Lefebvre (2015)*

These criteria include (i) a minimal solubility of the salt in water to increase the maximum concentration of the solution in the tank, and ultimately the energy storage density, and which is also an indicator of a strong interaction between the water molecules and the salt; (ii) a limited hydration degree, that is to say a low number of moles of water per mole of salt at the crystallization temperature, again to increase the maximum concentration of the solution in the tank; and (iii) a reasonable price for the anhydrous salt. This step results in a list of 11 "promising" salts, which are LiBr, LiCl, $ZnBr_2$, $ZnCl_2$, KHCOO, KCH_3COO, KSCN, NaSCN, NH_4HCOO, NH_4CH_3COO and NH_4SCN. A second step consists of studying the thermicity of the phenomena involved and the vapor pressures of the salt solutions. It shows that the dissolutions of the various salts in water as well as the dilutions of these salts are exothermic, except for NH_4HCOO, NH_4CH_3COO and NH_4SCN, where at least one of these phenomena is endothermic. The measurement of the vapor pressure of the solutions at different concentrations indicates a high vapor pressure value for KSCN and NaSCN, which makes them not very favourable for thermal storage by absorption. Finally, six salts underwent a more in-depth analysis, including the determination of the quantity of salt required to store 5,500 kWh and 1,200 kWh. The material costs, as well as the associated storage volumes, are also determined.

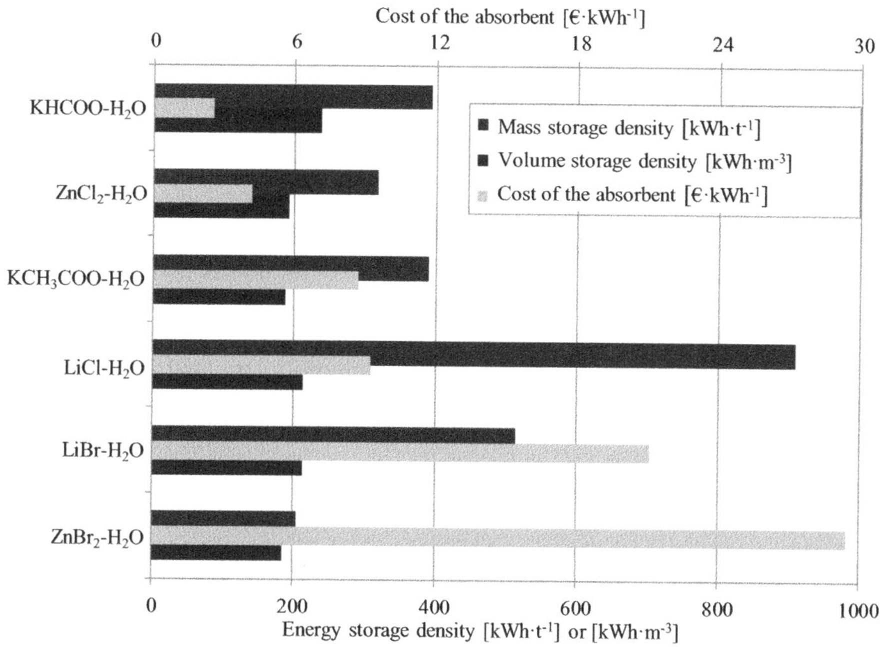

Figure 1.14. *Comparison elements of the main pairs with potential for thermal storage by absorption in the study by Lefebvre (2015). For a color version of this figure, see www.iste.co.uk/lepierres/heat2.zip*

It appears that LiBr and LiCl offer the best volume or mass energy storage densities, but are very expensive (Figure 1.14), meaning that it is unlikely that they could be considered for any commercial applications. It should be noted that, unlike the previous analysis, whose results are presented in Table 1.2, the current analysis assumes that the purchase price of LiBr (10,600 $€\cdot t^{-1}$) is higher than that of the LiCl (8,400 $€\cdot t^{-1}$). Conversely, $ZnBr_2$ and $ZnCl_2$ have a low volumetric energy density, requiring at least 9.4 m^3 of storage volume (water + solution) to store the 1,200 kWh to be released at 25°C under an evaporator at 10°C; their total cost is also high. Finally, it is KHCOO which offers a cost advantage and was therefore selected, despite its comparatively low thermal storage density (Figure 1.14).

1.5.2.2. *Physical adsorption*

According to the classification of the IUPAC (International Union of Pure and Applied Chemistry 2014), adsorbents can be microporous (pore size < 2 nm), mesoporous (pore size: 2–50 nm), or macroporous (pore size > 50 nm). Currently, silica gel and zeolite are the two main physical adsorbents considered for thermal storage processes by physical adsorption. This is already the case for adsorption-based cooling processes, where a study (Mugnier 2013) listing commercial machines indicates that out of the 12 machines surveyed, three operated with silica gel–H_2O, and the other nine used zeolite–H_2O. The main difference between zeolite and silica gel is that zeolite consists of microcrystals and, therefore, has a uniform pore size, while silica gel is an amorphous material with a pore size distribution ranging from micropores to macropores (Table 1.3). There are also aluminophosphates (Table 1.4), which are examined to a lesser extent for heat storage, as well as other adsorbents, including activated carbon, more commonly studied for use as a porous matrix for salts in thermochemical heat storage.

1.5.2.2.1. Silica gel

Silica gel is an amorphous material (Table 1.3), consisting of 99% silica (SiO_2) and hydroxyl groupings (OH), which makes it polar and gives it a greater affinity for polar sorbates, such as water and alcohols. The most widely used silica gels in sorption processes, including in sorption heat storage processes, are type A (from Fuji Davison) and type RD (regular diameter).

Properties	Zeolites	Silica gel
Chemical composition	$M^{+}_{x/n}$ $[(AlO_2^{-})_x(IfO_2)_y]\cdot zH_2O$ where M can be cations like Na^+, K^+, CA^{2+}, Li^+	SiO_2 Chemically inert, no metals or cations present
Chemical stability	Chemically stable in a basic medium, neutral, or slightly acidic	Chemically stable in a basic, neutral, or slightly acidic environment
Mechanical stability	The presence of liquid water can have an undesirable effect on mechanical stability	Gels with narrow pores are unstable when used with liquid water
Pore size	Pores with fixed or homogeneous sizes (crystalline structure) Zeolite A: ~3 Å or ~4 Å or ~5 Å Zeolite X: ~8 Å	Pore size distribution (amorphous structure) between 5 Å and 300 Å Average pore size of the gel with narrow pores: 20 Å Average pore size of the gel with large pores: 110 Å

Table 1.3. *Comparison of zeolites and silica gels (W.R. Grace & Co.-Conn 2010)*

Properties	Zeolites	Silica gel
Specific surface area ($m^2 \cdot g^{-1}$)	800	800 (narrow pore gels) 400 (wide pore gels)
Effective pores ($cm^3 \cdot g^{-1}$)	0.25–0.3	0.4 (narrow pore gels) 1.2 (wide pore gels)
Thermal conductivity ($W \cdot m^{-1} \cdot K^{-1}$)	0.12 (balls)	0.14–0.2
Heat of adsorption for water ($kJ \cdot kg^{-1}$ of water)	4,187	2,512
Specific heat capacity ($kJ \cdot kg^{-1} \cdot K^{-1}$)	0.96	0.92

Table 1.3. *Comparison of zeolites and silica gels (W.R. Grace & Co.-Conn 2010) (continued)*

The thermophysical properties of these two materials are similar, but the RD type has a larger specific surface area and higher thermal conductivity, which is enough to create a significant difference in their adsorption capacity (Wang et al. 2004). Adokou (2020) provides an inventory of properties, particularly adsorption properties, for these two materials, as well as for many other silica gels, including types 3A, A++, A5BW, RD 2060, RD 2560, grade 40 and 127B.

Basically, silica gel is inert and non-toxic. It only poses a health risk when it is saturated with a humidity indicator, such as cobalt chloride, which is a cancer-causing chemical compound (Rao et al. 2006).

Composites	Inorganic nanoporous materials	Polymers
Silica gel Mesoporous silicates Materials made from natural rocks Carbonaceous materials	Zeolites (aluminosilicates) Aluminophosphates Aerogels	Porous organometallic materials (MOF) Polyelectrolytes

Table 1.4. *Classification of physical adsorbents (Zheng et al. 2014)*

1.5.2.2.2. Zeolite

Zeolites (or zeolite) are aluminosilicates which have the general formula $M^+_{x/n}[(AlO_2^-)_x(SiO_2)_y]\cdot zH_2O$ (or $M_{x/n}[(AlO_2)_x(SiO_2)_y]\cdot zH_2O$), where M^+ represents the compensating cation and n represents its valence.

Zeolite has a crystalline structure with a determined or uniform pore size, unlike silica gel, which is amorphous with a pore size distribution. To achieve zeolites of a suitable size for applications, typically in the millimetre range, small-sized zeolite microparticles are agglomerated with binders to ensure good porosity and sufficient mechanical strength (Figure 1.15(a)). This results in a material with various shapes (Figure 1.15(b)) and dual porosity: micropores within the crystals and macropores between the crystals, meaning in the binder.

Zeolites can be natural or synthetic; there are currently about 48 natural zeolites and nearly 200 synthetic zeolites. However, few of them have been of interest for thermal storage applications; the ones that are of interest are types A and X zeolites. The main difference between types A and X zeolites is the pore size (Tables 1.5 and 1.6). Each zeolite crystal structure is represented by three letters, such as FAU, LTA, MFI, MEL, FER, LTL, SOO, etc., using the guidelines set out by the Structure Commission of the International Zeolite Association (IZA) (Baerlocher et al. 2007). This nomenclature is based on the ratio between the number of silica atoms and the number of aluminum atoms (Si/Al or y/x ratio) that they contain and on the number of rings formed by the pores.

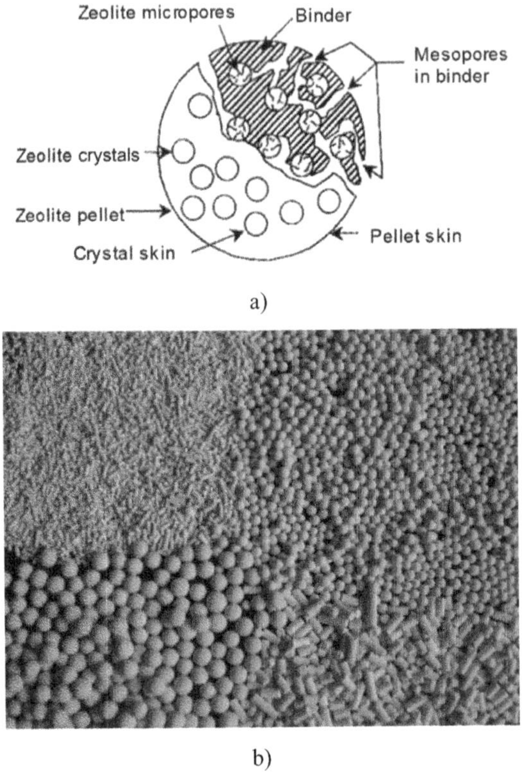

Figure 1.15. *(a) Structure of a zeolite grain (Sircar and Myers 2003); (b) zeolite grains in various forms (photos: courtesy of Pingxiang XINTAO Chemical Packing Co., Ltd.)*

For example, a type A zeolite, which does not have a natural counterpart (Dyer 2001), has a Si/Al ratio between 1 and 1.5 and forms rings with eight oxygen ions. The 3 Å diameter type A zeolite is called zeolite 3A

(Table 1.6). The one with a 5 Å diameter is called zeolite 5A. The MSX (medium silica X) zeolite has a crystalline structure corresponding to the natural faujasite (FAU) zeolite structure, with a Si/Al ratio between 1.05 and 1.15 (Nicolas et al. 2015). Zeolite 13X is also called NaX, with the Na+ cation being predominant in its structure (Table 1.6). Readers interested in nomenclature can refer to general works such as Payra and Dutta (2003).

	Type A	Type X
Chemical composition	$Na_{12}[(AlO_2)_{12}(SiO_2)_{12}]\cdot 27H_2O$	$Na_{86}[(AlO_2)_{86}(SiO_2)_{106}]\cdot 264H_2O$
Pore diameter (nm)	0.3–0.4	0.75 (NaX) 0.10 (CaX)
Specific internal surface area ($m^2 \cdot g^{-1}$)	800–1,000	800–1,000
Specific heat capacity ($kJ \cdot kg^{-1} \cdot K^{-1}$)	0.8–0.9	0.8–0.9
Thermal conductivity ($W \cdot m^{-1} \cdot K^{-1}$)	0.58	0.58
Bud bed density ($kg \cdot m^{-3}$)	750	700

Table 1.5. *Characteristics of the different type A and X zeolites*

Zeolite family	Structure of the corresponding natural zeolite	Si/Al ratio	Type of zeolites (examples)	Compensator cation	Pore diameter (Å)
Type A	Sodalite	1	Type 3A	K^+	3
			Type 4A	Na^+	4
Type X	Faujasite	1–1.5	Type 13X	NA^+ majority	~8
		1–1.5	Type 10X	Ca^{2+} majority	10
Type Y	Faujasite	1.5–3	Type Y	Na^+, K^+, Ca^{2+}	7
		>3	Type USY (Ultra Stable Y)	Na^+, K^+, Ca^{2+}	14

Table 1.6. *Classification of type A, X and Y zeolites (Dyer 2001; Sircar and Myers 2003; Hongois 2011)*

1.5.2.2.3. Other physical adsorbents

As previously presented (Table 1.4), there are other physical adsorbents considered for thermal energy storage applications by sorption. However, it should be noted that few authors have focused on them so far. This includes aluminophosphates, obtained by substituting phosphorus for silicon in the synthesis of zeolites, which are aluminosilicates.

The main advantage of aluminophosphates over their "cousins", the zeolites, is that they offer lower charging temperatures than zeolites, thanks in particular to the S-shape of their isotherms, which characterize type V adsorbents (Aristov 2007; Zheng et al. 2014). They also offer a higher energy storage density than zeolites. However, they are more expensive.

In this family, there are microporous aluminophosphates (AlPO), silicoaluminophosphates (SAPO) and metallic aluminophosphates (MeAPO), such as ferroaluminophosphates (FAPO). Various nomenclatures exist, including those of Union Carbide and IZA, leading to different designations for the same material, and each author or institute sometimes associates its name (MCM-n at Mobil, VPI-n at Virginia Polytechnic Institute, etc.) (Hafida 2008). For example, FAM-Z01 (Functional Adsorbent Material Zeolite 01) refers to FAPO-5, while FAM-Z02 refers to SAPO-34, and FAM-Z05 refers to FAPO-34. Readers interested in this nomenclature can refer to a bibliographic synthesis proposed by Hafida (2008).

There are also polymers (Table 1.4), including porous metal-organic frameworks (MOF) and polyelectrolytes. They have an adsorption capacity of over 80% of their weight with flexibility allowing the attainment of desired sorption isotherms (Zheng et al. 2014). MOFs are composed of different metal ions and organic ligands to form structures in one, two or three dimensions (Batten et al. 2013). They have an excellent water adsorption capacity but exhibit significant hydrophobic lengths at low pressures (Teo et al. 2017). Analyses have shown that MOFs degrade after several dozen cycles (Zheng et al. 2014).

1.5.2.2.4. Use of physical adsorbents in sorption heat storage processes

In the early 2000s, several researchers conducted studies on long-term heat storage using physisorbents (Table 1.7), which are deemed to be more stable than materials involving chemical reactions. Emphasis was placed on silica gels and zeolites, with some large-scale demonstration projects. However, these adsorbent materials proved to be inefficient due to low achieved energy storage densities (Table 1.7), low observed discharging temperatures, especially with silica gel, and charging constraints requiring a high source temperature, especially with zeolites.

Pair	Background and conditions of the study	Performances observed	References
Silica gel-H_2O	Closed system Experimental study of the material and theoretical analyses	Silica gels do not offer any potential for SSCS that would be of interest	(Bales 2008; Gantenbein 2008; Gantenbein et al. 2022)
Silica gel-H_2O	20.4 m^2 of flat plate collectors with DHW Charging: 88°C; Discharging: 42°C Prototypes of HYDES, then MODESTORE	HYDES prototype: 125 $kWh \cdot m^{-3}$ MODESTORE material: 50 $kWh \cdot m^{-3}$	(Gartler et al. 2004; Jähnig et al. 2006; Hadorn et al. 2007; Hauer 2007a; Bales et al. 2008)
Zeolite 13X–H_2O	Charging: 130°C; Discharging: 65°C Storage system for a district heating network	Heating: 124 $kWh \cdot m^{-3}$ Cooling: 100 $kWh \cdot m^{-3}$	(Hauer 2002b, 2007a)
	Charging: 180°C; Discharging: 55°C Closed system Prototype	Material: 180 $kWh \cdot m^{-3}$	(Bales et al. 2008)
	Charging: 180°C; discharging: 20°C	Material: 163 $kWh \cdot m^{-3}$	(Weber et al. 2016; Krese et al. 2018)
	Charging: 180°C; discharging: 57°C	Material: 180 $kWh \cdot m^{-3}$	(Stutz et al. 2017)
Zeolite 13XBF–H_2O	Charging: 190°C; discharging: 28°C T_{ev}: 10°C	Material: 198 $kWh \cdot m^{-3}$	(Gaeini et al. 2017)

Table 1.7. *Data from various sorption heat storage studies on physical adsorbents*

Pair	Background and conditions of the study	Performances observed	References
Zeolite 13X + activated alumina–H_2O	Charging: 250°C Tests with 55g of the AA13X material	Material: 200 $kWh \cdot m^{-3}$	(Dicaire and Tezel 2009, 2011)
Zeolite MSX–H_2O	Charging: 230°C; discharging: 50–60°C 50 kg of material in an open reactor	Material: 154 $kWh \cdot m^{-3}$	(Zettl et al. 2014)
Zeolite 4A–H_2O	Charging: 180°C; discharging: ≈ 35°C MONOSORP prototype	120 $kWh \cdot m^{-3}$	(Bales 2005; Kerskes 2006; Hadorn et al. 2007; Bales 2008)
Zeolite 4A–H_2O	Charging: 230°C; Discharging: 50–60°C 53 kg of material in an open reactor	Material: 148 $kWh \cdot m^{-3}$	(Zettl et al. 2014)
Zeolite 5A–H_2O	Charging: 103°C; Discharging: 51°C Closed system	Material: 47 $kWh \cdot m^{-3}$	(Scapino et al. 2017)
	Charging: 180°C; Discharging: 29°C	Material: 83 $kWh \cdot m^{-3}$	(Krese et al. 2018)
APO-n–H_2O	Charging: 95–140°C; Discharging: 40°C	Material: 240 $kWh \cdot m^{-3}$	–

Table 1.7. *Data from various sorption heat storage studies on physical adsorbents (continued)*

After some studies on physical adsorbents, including zeolites and silica gel, for thermal storage, the low performances led most researchers to consider the use of these materials only as a support or matrix for other materials with higher thermal storage density, especially salt hydrates, as discussed in section 1.5.2.4.

However, these adsorbent materials have proven to perform poorly because of the low storage energy densities that have been achieved (Table 1.7), the low discharging temperature levels observed (in particular with silica gel) and the charging constraints requiring a high source temperature, particularly with zeolites. After several studies of physical adsorbents for thermal storage, particularly zeolites and silica gel, these outcomes have led most researchers to consider the use of these materials as merely a support or supplement to other materials with a higher thermal storage density, particularly salt hydrates, as discussed in section 1.5.2.4.

At this point, it is worth making note of a reflection made by Aristov (2007) on choosing an adsorbent for energy applications. It is just as fast to select as it is to conceive of the material or materials that may be suitable for a given application based on the shape of the isotherm and the temperature conditions at the limits of the process taken from analytical relationships. According to this analysis, the materials whose sorption isotherms are S-shaped (type V and especially type VI, according to the IUPAC classification) are the materials that are best suited for thermal storage, a consideration which would make zeolites (type I) unattractive (Aristov 2007).

1.5.2.3. *Chemisorption*

The salt hydrates constitute the main class of materials used in heat storage by chemisorption, at least for low-temperature applications, meaning heat levels up to 100–200°C. Beyond 200°C, it is primarily oxide/hydroxide pairs that are employed.

1.5.2.3.1. Hydrated salts for thermochemical heat storage at low temperature

The analysis of the reaction enthalpies of salt hydrates shows that the release of one mole of water is accompanied by an absorption of approximately 55 kJ of heat when there is no particular modification of the salt structure, a value that can reach 74 kJ in certain cases (N'Tsoukpoe et al. 2014b). It follows that salt hydrates capable of releasing a large number of moles of water under operating conditions potentially offer a significant molar energy storage density. Since the volumetric energy storage density is related to the molar energy storage density, it follows that a salt releasing a large number of moles of water potentially provides a high volumetric energy storage density.

One of the first systematic studies on the selection of salt hydrates for low-temperature thermochemical heat storage was conducted by N'Tsoukpoe et al. (2014b). It is a study of the potential of the family of salt hydrates, taking into account the requirements of the operating conditions of the envisaged system. The approach is based on a three-step process, the essential elements of which are presented in Figure 1.16. This systematic approach has been taken up by other authors, adapted, or further developed for the selection of salt hydrates for thermochemical heat storage (Donkers et al. 2017) or the valorization of the thermal potential of industrial waste heat (Richter et al. 2018).

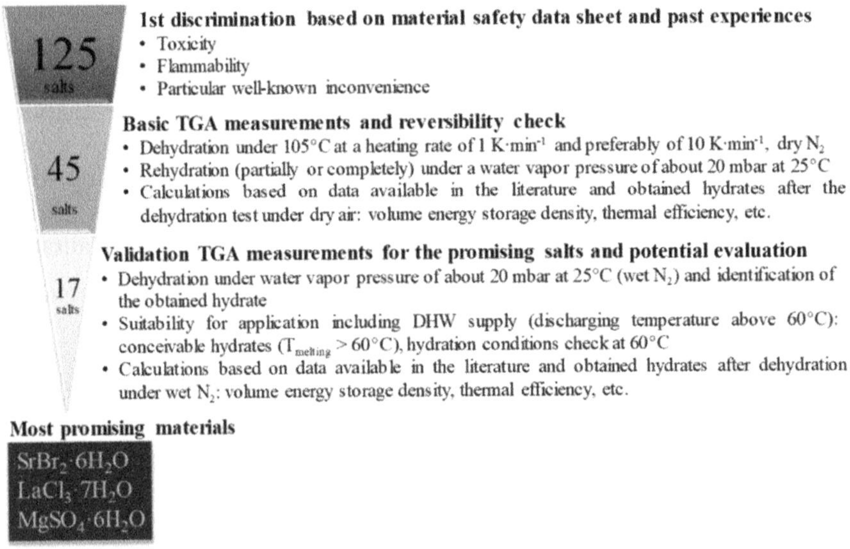

Figure 1.16. *Three-step approach for assessing the potential of thermochemical heat storage materials*

A list of 125 salt hydrates, which can be found in N'Tsoukpoe et al. (2014b), was established from chemistry databases. Initially, materials were eliminated based on general criteria, such as toxicity or flammability, the risk of explosion, or any particular danger or known problem based on past experiences. Materials exhibiting oxidizing behaviour were also rejected due to issues of cyclability, problems with the reversibility of the reaction after a certain number of cycles having been observed for these materials in previous tests. Only 45 materials (Table 1.8) passed through the filter of this first discrimination step and were thus retained for basic thermogravimetric analyses under dry nitrogen (Figure 1.16). The second and third steps combine theoretical calculations and experimental measurements based on TGA coupled with DSC.

In the second step, the authors conducted TGA/DSC analyses and checked for reversibility of the reactions. The purpose of these basic measurements is to verify the dehydration of the material in the temperature range below 105°C and the possible reversibility of the corresponding reaction with TGA/DSC. Among the 45 salts, 17 showed partial or complete rehydration under a vapor pressure of about 20 mbar at 25°C (Table 1.8). Using data available in the literature, the performance of these 17 salts was evaluated considering the hydrates formed during TGA. Figure 1.17 presents a synthesis of some results obtained, including volumetric energy densities and thermal efficiencies, as defined in section 1.3.2.2. The theoretical effective thermal efficiency η is calculated here by neglecting sensible heats; therefore, it reduces to the reaction enthalpy reduced by the heat of evaporation divided by the reaction enthalpy: this is the part of heat actually stored in the system as chemical potential (bond energy). It immediately appears that the theoretical effective thermal efficiency η of a conventional thermochemical heat storage unit based on salt hydrates below a maximum dehydration temperature of 105°C is relatively low (between 20% and 40%), considering only the heat effectively stored in the system as chemical potential. Two bromides offer the highest efficiencies: 39% for $MgBr_2 \cdot 6H_2O$ and 34% for $SrBr_2 \cdot 6H_2O$. In general, halogenated elements show the best performances (Figure 1.17). There are only three salts ($Zn(NO_3)_2 \cdot 6H_2O$, $CaBr_2 \cdot 6H_2O$ and $SrBr_2 \cdot 6H_2O$) that could provide an effective volumetric energy density of material greater than 200 $kWh \cdot m^{-3}$. A fourth one ($MgCl_2 \cdot 6H_2O$) is added to the previous list if aiming for an effective reduced volumetric energy density of material greater than 100 $kWh \cdot m^{-3}$. It has been reported that this material, often cited as interesting for thermochemical heat storage, cannot be completely dehydrated ($\rightarrow MgCl_2$) solely by heat input (Sharma 2002) and tends to chemically degrade with the formation of HCl (Ferchaud 2016; Sögütoglu et al. 2018).

Reversible (17)		Not promising (28)	
$Al_2(SO_4)_3 \cdot 18H_2O$	$SrBr_2 \cdot 6H_2O$	$AlNH_4(SO_4)_2 \cdot 12H_2O$	$LiBO_2 \cdot xH_2O$
$CaBr_2 \cdot 6H_2O$	$SrCl_2 \cdot 6H_2O$	$BaSO_4 \cdot xH_2O$	$Li(OH) \cdot H_2O$
$CaCl_2 \cdot 6H_2O$	$Zn(NO_3)_2 \cdot 6H_2O$	$Ca(CH_3COO)_2 \cdot xH_2O$	$Li_2SO_4 \cdot H_2O$
$Ce(SO_4)_2 \cdot 4H_2O$		$CaF_2 \cdot xH_2O$	$Mg_3(PO_4)_2 \cdot 8H_2O$
$K_2CO_3 \cdot 1.5H_2O$		$CaSO_4 \cdot 2H_2O$	$NaNO_3 \cdot xH_2O$
$KOH \cdot 2H_2O$		$Cs_2SO_4 \cdot xH_2O$	$Na_2B_4O_7 \cdot 10H_2O$
$LaCl_3 \cdot 7H_2O$		$CsCl \cdot xH_2O$	$NaCH_3COO \cdot 3H_2O$
$La(NO_3)_3 \cdot 6H_2O$		$Fe(NO_3)_3 \cdot 9H_2O$	$Na_2O_4W \cdot 2H_2O$
$LiCl \cdot H_2O$		$K_4[Fe(CN_6)] \cdot 3H_2O$	$Na_2SO_4 \cdot 10H_2O$
$LiNO_3 \cdot 3H_2O$		$K_2OTiO_2 \cdot xH_2O$	$NaHCO_3 \cdot xH_2O$
$MgBr_2 \cdot 6H_2O$		$K_2SO_4 \cdot xH_2O$	$NiNO_3 \cdot 6H_2O$
$MgCl_2 \cdot 6H_2O$		$KAl(SO_4)_2 \cdot 12H_2O$	$Sr(NO_3)_2 \cdot xH_2O$
$MgSO_4 \cdot 7H_2O$		$KBr \cdot xH_2O$	$TiO(SO_4) \cdot xH_2O$
$Na_2S_2O_3 \cdot 5H_2O$		$KCl \cdot xH_2O$	$ZnSO_4 \cdot 7H_2O$

Table 1.8. *Result of the basic thermogravimetric analysis: 17 reversible reactions and 28 non-promising reactions*

Sodium sulfide hydrates ($Na_2S \cdot xH_2O$) are also shown in Figure 1.17, even though they were eliminated in the preliminary selection phase due to their strong toxicity and corrosiveness. These salts are known for their high corrosivity and require relatively severe (low) operating pressures with other issues, such as the formation of H_2S and H_2 vapor (de Jong et al. 2014); it has also been highlighted that in the presence of CO_2 – hence atmospheric air – these materials are entirely unstable and lose their thermochemical qualities (Sögütoglu et al. 2018). However, they offer a high volumetric energy storage density (Figure 1.17), which decreases rapidly with cycles in the presence of CO_2. To use these materials, a pure substance in a hermetically sealed reactor would be required, increasing costs (Sögütoglu et al. 2018).

Since the melting temperature of $Na_2S \cdot 9H_2O$ is 49°C, only $Na_2S \cdot 5H_2O$ could be considered for applications requiring domestic hot water (DHW) production if no techniques are planned to manage the salt's melting.

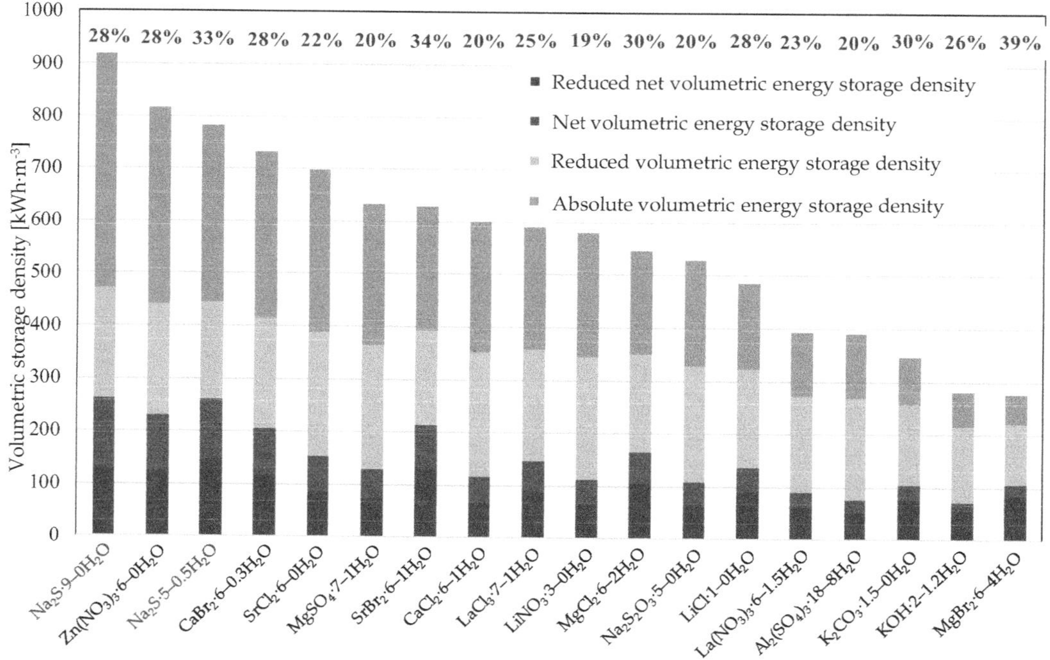

Figure 1.17. *The volumetric energy storage densities and the effective thermal efficiencies η for each of the salts that provide a reversible reaction after dehydration at 105°C in a dry atmosphere. The numbers above the histograms correspond to the thermal efficiency η. For example, for the reaction* $CaCl_2{\cdot}6H_2O + \Delta h_r \rightarrow CaCl_2{\cdot}H_2O + 5H_2O$, $\eta = 20\%$*; the absolute volumetric energy density is* $D_V = 601\ kWh{\cdot}m^{-3}$*, while the reduced effective volumetric energy storage density is* $D^*_{V,net} = 69\ kWh{\cdot}m^{-3}$*. For a color version of this figure, see www.iste.co.uk/lepierres/heat2.zip*

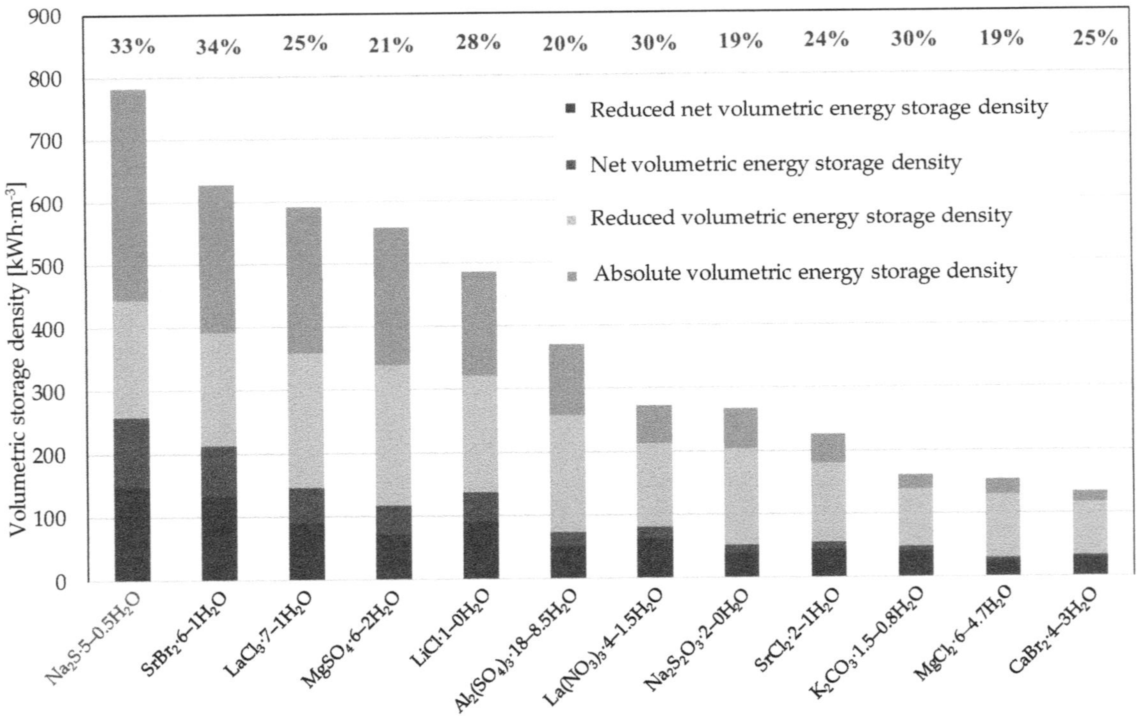

Figure 1.18. *Various volumetric energy densities and thermal efficiency for 11 of the 17 salts offering a reversible reaction after dehydration at 105°C under humid conditions (21 mbar water vapor pressure). For a color version of this figure, see www.iste.co.uk/lepierres/heat2.zip*

Third, a TGA to confirm the dehydration in a humid atmosphere and hydration at 60°C are carried out. The previous 17 materials are now subjected to a TGA under an atmosphere with a water vapor pressure of 21 mbar (equilibrium pressure of the water at 18°C) at 105°C, at a heating rate of 1 $K \cdot min^{-1}$. Only the hydrates with a melting point higher than 60°C are considered at this stage, since the problem of the melting of the salts is not one that is specifically intended to be addressed. The hydrates obtained at the end of the dehydration are then no longer necessarily the same since the atmosphere is no longer dry, reflecting real operating conditions, especially for closed systems. The obtained hydrates, as well the volumetric energy storage densities, are presented in Figure 1.18. There are five salts that have been eliminated at this stage.

According to the data presented in Figure 1.18, the top three salt hydrates are $SrBr_2 \cdot 6H_2O$, $LaCl_3 \cdot 7H_2O$ and $MgSO_4 \cdot 6H_2O$, excluding sodium sulfide, which was eliminated in a previous stage. $SrBr_2 \cdot 6H_2O$ offers the highest values for volumetric energy storage density and thermal efficiency.

The equilibria of the $MgSO_4 \cdot xH_2O$ system are complex, and they are too close to the pure water equilibrium curve. A minimum water vapor pressure of about 150 mbar (equivalent to 54°C at the evaporator) is required to reach 60°C in the reactor. Therefore, this salt is challenging to use, but a better understanding of the $MgSO_4 \cdot xH_2O$ equilibrium system could potentially change this conclusion. $SrBr_2 \cdot 6H_2O$ has already been documented and studied at the prototype scale for various applications. An approximately 80°C temperature is sufficient for complete dehydration, transitioning from hexahydrate to monohydrate. A water vapor pressure greater than 40 mbar (corresponding to water saturation at a temperature of 29°C) is necessary to reach 60°C in the reactor. Although there is no available data on $LaCl_3 \cdot xH_2O$ systems, $LaCl_3 \cdot H_2O$ was subjected to hydration under a water vapor pressure of 21 mbar at 60°C. Under these conditions, hydration led to the formation of $LaCl_3 \cdot 3H_2O$. This salt, which has not been considered for thermochemical heat storage, appears as an interesting candidate for such applications. Further studies could explore its hydration under higher water vapor pressures.

For an application requiring 60°C for discharging temperature, $SrBr_2 \cdot 6H_2O$ and $LaCl_3 \cdot 7H_2O$ emerge as the most promising salts, ignoring their prices. However, the maximum effective reduced volumetric energy density, a criterion used in contexts demanding high efficiency and considering the market value of the involved heat, offered by these two salts

(133 kWh·m^{-3} and 89 kWh·m^{-3}, respectively), would likely fall short of the target, namely an effective reduced energy storage density typically set around 100 kWh·m^{-3} for the storage unit. Even for applications requiring a low discharging temperature, for example, 35°C, the achievable effective thermal storage density and thermal efficiency are relatively low. Despite the prevalence of salt hydrates as storage materials, most salts have led to technical performances below expectations, including a discharging temperature, thermal storage density or power density (kinetics) that is too low or insufficient, not to mention stability issues (cyclability).

In her doctoral thesis, Druske (2020) focused on mixtures of multiple salts obtained either by simple physical mixing or by dissolution followed by evaporation/crystallization. The aim was to improve the stability (cyclability) of materials and their energy storage density. Sixty-two mixtures were prepared using sulfates, chlorides and bromides of the following cations: Na^+, K^+, Li^+, Sr^{2+}, CA^{2+}, Mg^{2+}, Zn^{2+}, Fe^{2+}, Fe^{3+} and Al^{3+} (Table 1.9). It is noteworthy that almost all of these salts were individually considered in the study by N'Tsoukpoe et al. (2014b) mentioned earlier.

From this set, only four mixtures ($\{2MgCl_2 + KCl\}$, $\{2MgCl_2 + CaCl_2\}$, $\{5SrBr_2 + 8CaCl_2\}$ and $\{2ZnCl_2 + CaCl_2\}$) were found interesting (high mass energy storage density, stability of the mixture during cycling) following ATG/DSC measurements.

Further studies on a larger scale (recycling from 10–25 cycles of 20–100 g of material) show that only $\{2MgCl_2 + KCl\}$ and $\{2MgCl_2 + CaCl_2\}$ would be feasible for thermochemical heat storage. In the case of $\{2MgCl_2 + CaCl_2\}$, the formation of Mg(OH)Cl and HCl occurs when the temperature exceeds 110°C without affecting the energy storage density. This formation of hydrochloric acid is problematic for the use of the material in thermochemical heat storage. This decomposition is not observed when using $\{2MgCl_2 + KCl\}$ up to a temperature of 200°C.

Overall, in the study, KCl proves to be an interesting stabilizer, although it itself practically does not store heat, which negatively affects the energy storage density. The study did not clearly demonstrate that the mixture could offer a (significantly?) greater thermal storage density than the material originally having the highest energy storage density before the mixture.

Category	Mixtures
Sulfates	{$3Na_2SO_4 + K_2SO_4$}, {$Na_2SO_4 + K_2SO_4$}, {$Na_2SO_4 + 3K_2SO_4$}, {$2Na_2SO_4 + MgSO_4$}, {$7Na_2SO_4 + 4MgSO_4$}, {$2K_2SO_4 + MgSO_4$}, {$7Na_2SO_4 + 4ZnSO_4$}, {$10K_2SO_4 + 7ZnSO_4$}, {$3MgSO_4 + 2ZnSO_4$}, {$Na_2SO_4 + Fe_2 + SO_4$}, {$2Na_2SO_4 + Fe_2 + SO_4$}, {$9K_2SO_4 + 5Fe_2 + SO_4$}, {$2Fe_n(SO_4)_m + Al_2(SO_4)_3$}, {$2Na_2SO_4 + Al_2(SO_4)_3$}, {$17MgSO_4 + 3Al_2(SO_4)_3$}, {$MgSO_4 + Al_2(SO_4)_3$}
Chlorides	{$MgCl_2{\cdot}6H_2O$}, {$CaCl_2{\cdot}2H_2O$}, {$ZnCl_2$}, {KCl}, {$CaCl_2 + 2MgCl_2$}, {$CaCl_2 + MgCl_2$}, {$2CaCl_2 + MgCl_2$}, {$2CaCl_2 + ZnCl_2$}, {$CaCl_2 + ZnCl_2$}, {$CaCl_2 + 2ZnCl_2$}, {$2MgCl_2 + ZnCl_2$}, {$MgCl_2 + ZnCl_2$}, {$MgCl_2 + 2ZnCl_2$}, {$2MgCl_2 + KCl$}, {$MgCl_2 + KCl$}#1, {$MgCl_2 + KCl$}#2, {$MgCl_2 + 2KCl$}, {$2CaCl_2 + KCl$}, {$CaCl_2 + KCl$}, {$CaCl_2 + 2KCl$}, {$8ZnCl_2 + 7KCl$}, {$4ZnCl_2 + 7KCl$}, {$2ZnCl_2 + 7KCl$} {$MgCl_2 + CaCl_2 + ZnCl_2$}, {$MgCl_2 + CaCl_2 + 2KCl$}, {$MgCl_2 + ZnCl_2 + 2KCl$}, {$CaCl_2 + ZnCl_2 + 2KCl$}, {$MgCl_2 + CaCl_2 + ZnCl_2 + 3KCl$}
Bromides	{$SrBr_2{\cdot}6H_2O$}, {$NaBr$}, {KBr}, {$LiBr$}, {$MgBr_2{\cdot}6H_2O$}, {$CaBr_2{\cdot}xH_2O$}, {$2NaBr + SrBr_2$}, {$NaBr + SrBr_2$}, {$NaBr + 2SrBr_2$}, {$2KBr + SrBr_2$}, {$KBr + SrBr_2$}, {$KBr + 2SrBr_2$}, {$2LiBr + SrBr_2$}, {$LiBr + SrBr_2$}, {$LiBr + 2SrBr_2$}, {$2MgBr_2 + SrBr_2$}, {$MgBr_2 + SrBr_2$}, {$MgBr_2 + 2SrBr_2$}, {$2CaBr_2 + SrBr_2$}, {$CaBr_2 + SrBr_2$}, {$CaBr_2 + 2SrBr_2$}, {$5CaBr_2 + 4SrBr_2$}, {$5CaBr_2 + 8SrBr_2$}, {$5CaBr_2 + 16SrBr_2$}
Sulfates + Chlorides	{$16KCl + 3MgSO_4$}, {$ZnSO_4 + 36ZnCl_2$}
Sulfates + Bromides	{$4SrBr_2 + 3MgSO_4$}, {$2SrBr_2 + 3MgSO_4$}, {$SrBr_2 + 3MgSO_4$}
Chlorides + Bromides	{$2SrBr_2 + 5KCl$}, {$SrBr_2 + 5KCl$}, {$SrBr_2 + 10KCl$}, {$5SrBr_2 + 4CaCl_2$}, {$5SrBr_2 + 8CaCl_2$}, {$5SrBr_2 + 16CaCl_2$}

Table 1.9. *Complete group of salts and salt mixtures studied by Druske (2020)*

1.5.2.3.2. Materials for thermochemical heat storage at high temperature

Different categories of chemical reactions have been studied for thermochemical heat storage at medium and high temperatures (Table 1.10).

The system of $CaO/Ca(OH)_2$ is by far the one that has attracted the most attention in recent years among the primary systems studied for thermal storage by chemical reactions for temperatures ranging from about 200°C to 600°C, or even higher. By adjusting the water vapor pressure, one can easily modify or control the discharging temperature up to beyond 600°C (Liu et al. 2018). CaO, like MgO, is known to be readily available, non-toxic and inexpensive (Pardo et al. 2014; Liu et al. 2018; Humphries et al. 2019) and the reaction is easy to implement. One of the main challenges with this material remains the issue of particle agglomeration and its low thermal conductivity. Processes involving the use of ammonia are the most mature (Table 1.10), notably because the synthesis of ammonia from nitrogen and hydrogen, which are relatively abundant reactants, is a well-established industrial technique. As it is a gas-phase-only reaction, it exhibits excellent reversibility compared to solid–gas reactions, with both gases being well known and stable within the envisaged temperature range. Moreover, there is no need to store the fluids in separate containers since their densities are significantly different at room temperature and moderate pressure (Liu et al. 2018).

However, the harsh operating conditions, including high pressure and temperature, and safety requirements for gas storage contribute to the high operational cost of using this reaction for heat storage.

The achieved thermal storage energy density in this temperature range could also reach almost 900 $kWh{\cdot}m^{-3}$ (Table 1.10). However, thermal energy storage density is not necessarily a primary concern for industrial storage. In industrial settings, considerations such as space constraints and availability are different from those in residential areas. The primary concern is whether the reaction is technically feasible at an acceptable cost. Even safety constraints are more manageable here than in residential applications. For instance, ammonia storage is feasible in an industrial context, which is practically impossible for a domestic application, as explained in section 1.5.1.

Reaction category	Reaction	Operating conditions	Storage density
Hydroxides	$Mg(OH)_2 + \Delta H_r \leftrightarrow MgO + H_2O$ (solid–gas)	$T_c = 150°C$ $T_d = 100°C$	$388\ kWh \cdot m^{-3}$ $390\ kWh \cdot t^{-1}$
	$Ca(OH)_2 + \Delta H_r \leftrightarrow CaO + H_2O$ (solid–gas)	$T_c = 450°C$ $T_d = 25–400°C$	$437\ kWh \cdot m^{-3}$ $390\ kWh \cdot t^{-1}$
Hydrides	$MgH_2 + \Delta H_r \leftrightarrow Mg + H_2$ (solid–gas)	$T_c = 380°C$ $T_d = 230°C$	$580\ kWh \cdot m^{-3}$ $800\ kWh \cdot t^{-1}$
	$CaH_2 + \Delta H_r \leftrightarrow Ca + H_2$ (solid–gas)	$T_c = 380°C$ $T_d = 230°C$	$580\ kWh \cdot m^{-3}$ $1{,}071\ kWh \cdot t^{-1}$
Carbonates	$PbCO_3 + \Delta H_r \leftrightarrow PbO + CO_2$ (solid–gas)	$T_c = 450°C$ $T_d = 300°C$	$303\ kWh \cdot m^{-3}$ $90\ kWh \cdot t^{-1}$
	$CaCO_3 + \Delta H_r \leftrightarrow CaO + CO_2$ (solid–gas)	$T_c = 860°C$ $T_d = 880°C$	$692\ kWh \cdot m^{-3}$ $490\ kWh \cdot t^{-1}$
Ammoniacal solutions	$NH_4HSO_4 + \Delta H_r \leftrightarrow NH_3 + H_2O + SO_3$ (liquid–gas)	$T_c = 927°C$ $T_d = 427°C$	$860\ kWh \cdot m^{-3}$ $810\ kWh \cdot t^{-1}$
	$2NH_3 + \Delta H_r \leftrightarrow N_2 + 2H_2$ (gas)	$T_c = 450°C$ $P_c = 9–150$ bars $T_d = 450°C$ $P_d = 10–300$ bars	$745\ kWh \cdot m^{-3}$ $1{,}090\ kWh \cdot t^{-1}$

Table 1.10. *Reactions studied for high temperature thermochemical heat storage (Pardo et al. 2014; Humphries et al. 2019). The storage energy density is given relative to the first material in the left side of the equation*

Reaction category	Reaction	Operating conditions	Energy storage density
Redoxes	$2Co_3O_4 + \Delta H_r \leftrightarrow 6CoO + O_2$ (solid–gas)	$T = 870–900°C$	$295\ kWh \cdot m^{-3}$ $240\ kWh \cdot t^{-1}$
	$2BaO_2 + \Delta H_r \leftrightarrow 2BaO + O_2$ (solid–gas)	$T = 690–780°C$	$328\ kWh \cdot m^{-3}$ $130\ kWh \cdot t^{-1}$
Organics	$CH_4 + H_2O + \Delta H_r \leftrightarrow 3H_2 + CO$ (gas)	$T_c = 950°C$ $T_d = 530°C$	$7.8\ kWh \cdot m^{-3}$ $4{,}340\ kWh \cdot t^{-1}$
	$CH_4 + CO_2 + \Delta H_r \leftrightarrow 2H_2 + 2CO$ (gas)	$T_c = 950°C$ $T_d = 530°C$	$7.7\ kWh \cdot m^{-3}$ $4{,}280\ kWh \cdot t^{-1}$
	$C_6H_{12} + \Delta H_r \leftrightarrow C_6H_6 + 3H_2$ (liquid–gas)	$T_c = 317°C$ $T_d = 397°C$	$530\ kWh \cdot m^{-3}$ $680\ kWh \cdot t^{-1}$
SO_3	$2SO_3 + \Delta H_r \leftrightarrow 2SO_2 + O_2$ (liquid–gas)	$T_c = 800°C$ $T_d = 500–600°C$	$646\ kWh \cdot m^{-3}$ $340\ kWh \cdot t^{-1}$

Table 1.10. *Reactions studied for high-temperature thermochemical heat storage (Pardo et al. 2014; Humphries et al. 2019). The energy storage density is given relative to the first material on the left side of the equation (continued)*

1.5.2.4. *Coupled chemisorption and physisorption*

Due to various stability issues with materials, several research teams have shifted toward using matrices of physisorbents (typically zeolite or silica gel) that are more stable. These matrices are usually impregnated with salt hydrates. The volumetric energy storage density takes a hit, decreasing by approximately 60–70% (Hongois et al. 2011; Nonnen et al. 2016) for at least two reasons. First, these matrices have a lower volumetric storage energy density compared to the salts. Second, the amount of salt hydrates that can be impregnated to obtain a stable and sufficiently porous material is generally limited (usually 5–20%) (Hongois et al. 2011; Mette et al. 2013; Nonnen et al. 2016). Additionally, depending on operating conditions, the composite material may exhibit a lower energy storage density than each of the individual materials (Whiting et al. 2013; Nonnen et al. 2016). In such cases, the composite material becomes less advantageous, and it is often more practical to use the stable material alone, as seen with certain zeolites (Whiting et al. 2013; Metchueng Kamdem 2016). Zeolites can achieve discharging temperatures suitable for DHW needs, but the required charging temperature is generally higher, exceeding 150°C, and sometimes 250°C. Although it is possible to reduce the charging temperature, it comes at the expense of energy storage density (Lass-Seyoum et al. 2016).

The use of additive materials with no real sorption capacity, especially those intended for improving thermal or mass transfer or maintaining the stability of the active material, is discussed in section 1.6.

1.6. Introduction to the issue of heat and mass transfer in solid–gas storage materials

1.6.1. *Kinetics of the adsorption phenomenon or solid–gas reaction*

Let us consider a fire truck with a full water tank with a capacity of 6 m^3 arriving at a fire. If the opening of the hose connected to the tank can only deliver a flow rate of 15 L·h^{-1}, equivalent that of a shower head, there is little chance that the firefighters will be able to put out the flames in time, even though a 6 m^3 capacity would normally be sufficient to extinguish the fire in question. It is not enough to simply have a sufficient water storage that water must also be transferred at a sufficient flow rate. The same is true for thermal energy storage. It is not enough to store a large amount of heat in a small storage unit: it is also necessary to be able to recover the heat at a rate, or power, sufficient to meet the user's needs. A storage of 1,800 kWh of heat is of little use if, during the cold period when the building requires a power of 1 kW for the comfort of its occupants, the storage can only provide 100 W! A high specific power, meaning the power that can be exchanged with the material per unit mass or volume, is one of the key performance elements of a reactor and, consequently, of a thermochemical heat storage material. Specific power depends, among other things, on the kinetics of the reaction in play. The reaction kinetics in a solid–gas reactor will be presented in more detail in Chapter 3, which is dedicated to reactors. The aim here is to introduce the origin of transfer limitations at the material level and some approaches implemented to reduce these limitations and improve transfers.

Thermal energy storage reactors by sorption or chemical reaction involve coupled heat and mass transfers. Aside from the periods of temperature rise/fall of the material in the reactor, at the beginning or end of operation, any heat transfer in the material is accompanied by gas release/fixation as long as the temperature and pressure conditions allow the sorption phenomenon. The kinetics of the phenomenon in the reactor depends precisely on heat and mass transfers, as well as the *deviation from equilibrium*. The analyses presented in this chapter are generally under the thermodynamic equilibrium conditions of the considered pairs. However, at thermodynamic equilibrium, by definition, the reaction rate is zero, and nothing happens. For example, when two identical pieces of copper at the same temperature are in thermal contact, nothing happens, as the potential for transfer, here at the temperature, is the same. In contrast, when one of the two pieces is initially at a higher temperature than the other, heat transfer is observed between the two pieces until thermal equilibrium is reached. This transfer decreases, that is, it becomes less and less rapid, with the decrease in temperature difference between the two pieces. Similarly, in a thermochemical reactor, the reaction kinetics are faster when the deviation from equilibrium conditions, or deviation from equilibrium, is large. This can be, for example, a difference in pressure or temperature. Thus, a sufficiently high water vapor pressure for discharging (lower for charging) is necessary to achieve a sufficient reaction kinetics (Rambaud 2009; Michel 2012; Michel et al. 2012; Richter et al. 2018).

Adsorbents are used in the form of powders, pellets, granules or beads. These forms and sizes of the material will influence not only the energy storage density but also important aspects such as mass and heat transfer within the material. Adsorbents are, by definition, porous materials; they generally exhibit low thermal conductivity (Table 1.11). This low thermal conductivity, coupled with that of the gaseous sorbate filling the pores, especially at low pressure (Table 1.11), and the low thermal contacts between solid grains due to intergranular discontinuity/porosity, leads to a relatively low overall heat transfer in the reactor, primarily occurring through conduction across the material bed (Figure 1.19). Furthermore, the decomposition reaction, for example, dehydration of a salt hydrate, is accompanied by a decrease in the volume of the salt grain, while the synthesis reaction, for example, hydration, is accompanied by an increase in the volume of the salt grain. These volume variations lead to modifications, particularly those of the texture and porosity of the material, significantly affecting heat transfers within the material.

Mass transfer, or gas diffusion into the material, depends, among other factors, on permeability, which is a measure of the ability of a porous material to be "passed through" by a fluid. This mass transfer is generally poor in low-pressure processes due to the low pressure of the gaseous phase. One of the phenomena frequently observed in thermochemical reactors using chemisorbents, generally salt hydrates and hydroxides, is the decrease in the permeability of the material after a few cycles, that is, a succession of decomposition and synthesis reactions. This decrease in permeability is related to agglomeration of solid particles in the reactor. It may be just the formation of a surface crust that prevents vapor from diffusing and accessing the grains available under the crust. One can think of the formation of a crust on a pile of powdered sugar deposited in a relatively humid place. It can also be a mass uptake of the entire material. This is the example of quicklime powder deposited in a pile in a humid outdoor environment: one obtains a more or less hard block. Sometimes, the loss of permeability is the result of solidification following a more or less accidental melting of the material when its melting temperature is locally reached (see Box 1.1).

The loss of permeability can also be due to "overhydration": the salt or part of the salt, typically the surface layer, absorbs too much water, forming a paste or even a solution. The subsequent dehydration of this paste can lead to an increase in the compactness of the material. This happens with highly deliquescent materials, such as calcium chloride ($CaCl_2{\cdot}nH_2O$). In any case, regardless of the origin of the loss or reduction in permeability, the result is the same: the gas (sorbate) no longer accesses the entire solid material, and the reaction becomes impossible. This is one of the main limitations of using chemisorbents.

Material	Form	Thermal conductivity [W·m⁻¹·K⁻¹]
Zeolite	Solid grain	0.18–0.4
	Granular bed	0.03–0.15
	Granular bed (H_2O)	0.03–0.15
Activated carbon	Solid grain	0.54
	Granular bed	0.3–0.5
	Granular bed (methanol)	0.14–0.17
Silica gel	Solid grain	0.37–0.8
	Granular bed	0.05–0.20
	Granular bed (H_2O)	0.04–0.26
$CaCl_2$	Granular bed	0.1–0.2
$Ca(OH)_2$	Granular bed	0.1
Water vapor	Vapor (6–100 mbar)	0.016–0.020

Table 1.11. *Thermal conductivity of various materials (solid grains) and effective thermal conductivity of unconsolidated beds (granular beds) (N'Tsoukpoe et al. 2014a)*

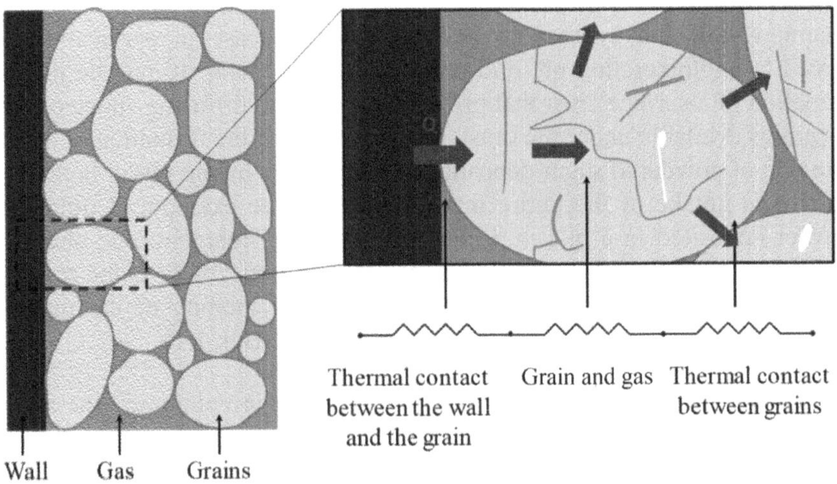

Figure 1.19. *Thermal resistances along the thermal transfer path in a porous bed with a heat exchanger (N'Tsoukpoe et al. 2014a). For a color version of this figure, see www.iste.co.uk/lepierres/heat2.zip*

$Na_2S \cdot 9H_2O$ is a salt that can release up to 9 moles of water per mole of salt, which means that in principle it offers a high volumetric energy storage density. This is also the case if we consider the reaction $Na_2S \cdot 9H_2O + \Delta H_r \leftrightarrow Na_2S + 9H_2O$, which offers an energy storage density of 924 $kWh \cdot m^{-3}$ for $Na_2S \cdot 9H_2O$, one of the highest among all salt hydrates (Figure 1.17). However, the melting temperature of $Na_2S \cdot 9H_2O$ is 49°C. It has been observed that when the rate of heating of the material in a reactor or in a calorimeter exceeds 1.6 $K \cdot min^{-1}$, the complete dehydration of the $Na_2S \cdot 9H_2O$ does not have time to occur before the temperature reaches 49°C (this temperature needs to be exceeded to obtain Na_2S under the operating conditions). Consequently, partial or almost total melting of the material occurs, accompanied by a modification of its structure that hinders or limits subsequent hydrations (de Boer et al. 2002). Thus, depending on the rate of heating or temperature rise of the material, the partial or total melting of the material occurs instead of dehydration. This can lead to the destruction of the structure of the material and result in particle agglomeration.

For use requiring the production of domestic hot water, only $Na_2S \cdot 5H_2O$ (with a melting temperature of 83°C) is considered if there are no planned techniques to manage the melting of the salt using the reaction $Na_2S \cdot 5H_2O + \Delta H_r \leftrightarrow Na_2S \cdot 0.5H_2O + 4.5H_2O$. Even in this case, the melting of the $Na_2S \cdot 5H_2O$ is observed under a heating rate of 1.6 $K \cdot min^{-1}$ when exceeding 83°C in the presence of water vapor.

During the hydration phase $Na_2S \cdot 2H_2O + 3H_2O \rightarrow Na_2S \cdot 5H_2O + \Delta H_r$, which can achieve a temperature of 60°C for a water evaporation temperature of about 10°C, the formation and fusion of $Na_2S \cdot 9H_2O$ can be observed, while all of $Na_2S \cdot 2H_2O$ has not yet become $Na_2S \cdot 5H_2O$.

All of these factors make the use of Na_2S hydrates challenging, as it requires specific operating conditions and precautions to harness their significant volumetric energy density for storage, not to mention other problems mentioned in section 1.5.2. Similar issues exist for other highly hydrated salts with a relatively low melting temperature, such as $AlNH_4(SO_4)_2 \cdot 12H_2O$ or $KAl(SO_4)_2 \cdot 12H_2$ (Galwey and Brown 1999; Marias et al. 2014; N'Tsoukpoe et al. 2014b).

Box 1.1. *Na_2S hydrates melting issues*

1.6.2. *Improvements to the characteristics of thermochemical heat storage materials*

According to Rambaud (2009), the agglomeration of salt grains occurs during synthesis reactions carried out under conditions close to thermodynamic equilibrium, especially when these are not far from the salt dissolution conditions (saturated solution). Consequently, a proper determination of the implementation conditions of the reaction could avoid the formation of agglomerates. However, researchers often opt for another approach: dispersing the salt in a more stable material (Table 1.12). One of the most commonly used matrices for this purpose is vermiculite. Others use more stable physisorbents, such as silica gel or zeolite (Figure 1.20), as matrices, allowing them to benefit from the sorption capacity of these materials.

The main technique used to immobilize the salt in the matrix pores is impregnation. It involves first drying the matrix to remove any absorbed water before impregnating the matrix particles in an aqueous solution of the salt (Gordeeva et al. 2013). Any excess solution that is not fixed in the matrix pores is removed either by filtration or by vacuum drying. The amount of salt that can be fixed in the matrix pores without encountering the formation of agglomerates again is relatively limited, significantly affecting the volumetric energy storage density.

In different studies (Hongois 2011; Marias 2015), the composites do not exhibit more than 20–40% of their absolute volumetric energy storage density, not forgetting that the bonding energy could also decrease due to the physisorbents (section 1.5.2.4). It is possible that some salt hydrates melt but regain their original structure, especially if they are inserted into another stable matrix, as seems to be the case with $CaCl_2$ confined in silica gel (Aristov et al. 1996). It is also observed that the material resulting from the impregnation of a salt (monovariant equilibrium) into the pores of a physisorbent (divariant equilibrium) can follow a monovariant equilibrium (Gordeeva et al. 2013), but this does not seem to always be the case (Hongois 2011).

As for improving heat transfer, various techniques are employed, such as adding metallic or carbon fibers to the material. In the case of carbon fibers, a consolidated material is typically produced, meaning a compact solid structure. This tends to reduce permeability, and the choice of matrix, as well as the implementation, is carried out to ensure and preserve minimal porosity. One of the most commonly used carbon materials is expanded natural graphite (ENG). It is a material with a very low apparent density (~3–4 $kg \cdot m^{-3}$), high intrinsic thermal conductivity (~200 $W \cdot m^{-1} \cdot K^{-1}$), significant porosity, and good elasticity (Stitou 2013). The resulting consolidated composite material has high porosity (60–90%) and high effective thermal conductivity (5–40 $W \cdot m^{-1} \cdot K^{-1}$), depending on the apparent density of ENG in the composite material (~50–200 $kg \cdot m^{-3}$). Increasing the amount of ENG thus increases thermal conductivity, favouring heat transfer in the material but reducing its permeability, which hinders mass transfer (Figure 1.21). The choice of the amount of ENG to implement depends on the requirements of the intended application.

For thermochemical heat storage, it is clear that the amount of ENG should be kept minimal to maximize thermal energy storage density. Because long-term thermochemical heat storage requires a significant amount of material, corresponding to a low average volumetric power requirement, some authors argue that the use of graphite in this context is not justified (Stitou 2013; Michel et al. 2014). This requirement for low volumetric power is indeed low if one considers that the entire material in the storage is subjected to the reaction simultaneously, as suggested by Michel et al. (2014): for a storage system involving 5–10 t of $SrBr_2 \cdot 6H_2O$, the authors estimate it at 0.3–0.8 $W \cdot kg^{-1}$ by dividing the required thermal power (3–4 kW) by the total mass of the storage material. However, the assumption of the simultaneous reaction of all storage material, that is, several tons of salt, is generally not adopted in the literature, explaining the use of carbon materials, even for long-term thermal energy storage (Table 1.12).

The use of composite materials is becoming increasingly systematic in the case of solid sorbents, with porous matrices, especially zeolites, being employed to disperse and stabilize the medium while taking advantage of their storage capacity. This provides the opportunity to lower the charging temperature while modifying the sorption characteristics to adapt to operating conditions. The main drawback of these materials remains the reduction in volumetric energy storage density and the decrease in the temperature of heat release for the same water vapor pressure.

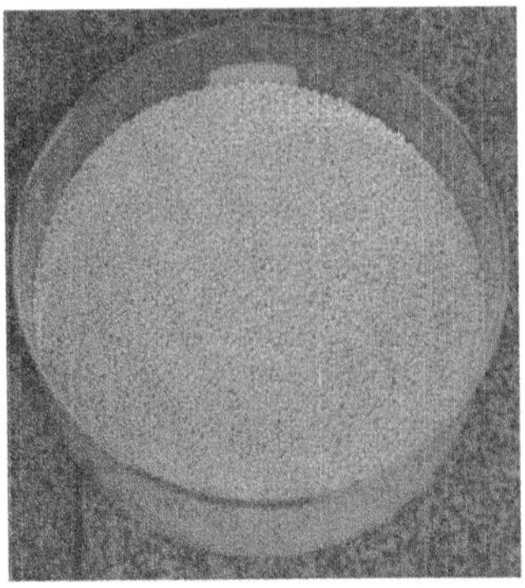

Zeolite

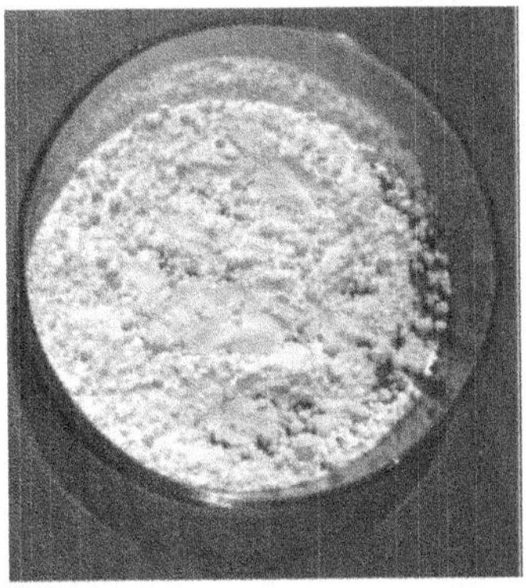

$MgSO_4$

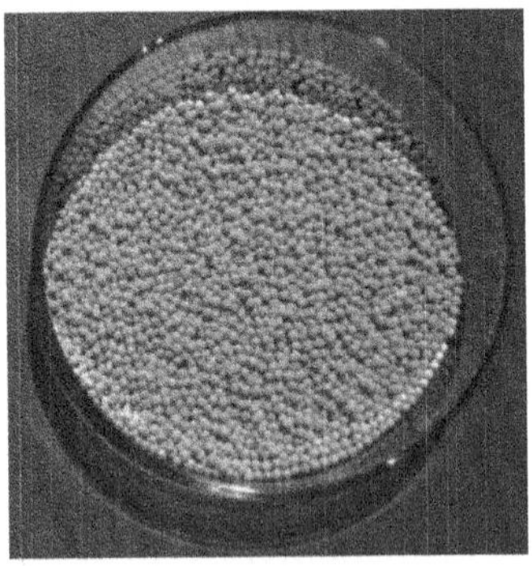

Composite of Zeolite-$MgSO_4$

Figure 1.20. *Zeolite 13X, $MgSO_4$ and zeolite-$MgSO_4$ composite with a mass fraction of 15% of $MgSO_4$ (Hongois 2011). For a color version of this figure, see www.iste.co.uk/lepierres/heat2.zip*

Matrix	Salt	Comments	References
Attapulgite	$CaCl_2$ (30 m%)	–	(Jänchen et al. 2005)
Attapulgite	$MgSO_4$, $MgSO_4$ + $MgCl_2$	Mixtures of salt in different proportions, use of impregnation techniques to increase the energy storage density and heat transfers	(Posern and Kaps 2010)
Aluminosilicate	$CaCl_2$	–	(Jänchen et al. 2004)
Bentonite	$CaCl_2$ (40 m%)	Impregnation, maximum energy storage density of the composite material: 185 $kWh \cdot m^{-3}$	(Kerskes et al. 2010)
Expanded natural graphite (ENG)	$CaCl_2$	–	(Tian et al. 2012)
Activated carbon, silica gel, zeolite, vermiculite	$CaCl_2$, $MgSO_4$, $Ca(NO_3)_2$, $Li(NO_3)_2$, LiBr	–	(Casey et al. 2014)
Activated carbon	$CaCl_2$	–	(Casey et al. 2014)
Activated carbon fiber	$CaCl_2$	–	(Druske et al. 2014)
γ-Alumina 'IK-02-200'	$CaCl_2$	–	(Gordeeva et al. 1998)
Carbon sibunit	LiBr	–	(Gordeeva et al. 2002)
Activated carbon/silica gel	$CaCl_2$	–	(Tso and Chao 2012)
Alumina	$CaCl_2$	–	(Tanashev et al. 2013)
Silica-alumina	$CaCl_2$, $Ba(OH)_2$, $LiNO_3$	–	(Jabbari-Hichri et al. 2015)
Mesoporous silica gel	$CaCl_2$	–	(Aristov et al. 1996)
Silica gel	$CaCl_2$, $MgSO_4$, $Ca(NO_3)_2$, $Li(NO_3)_2$, LiBr	–	(Casey et al. 2014)
Microporous silica gel	$Ca(NO_3)_2$	–	(Simonova et al. 2009)

Matrix	Salt	Comments	References
KSK silica gel	$CaCl_2$, $MgCl_2$, LiBr	–	(Tanashev et al. 2013)
Silica gel	$CaCl_2$	–	(Wu et al. 2007)
Silica gel	LiCl	–	(Gordeeva et al. 2009)
Silica gel	$MgSO_4$	–	(Aristov and Gordeeva 2009)
Silica gel	$NaSO_4$	–	(Aristov and Gordeeva 2009)
Silica gel	$Ca(NO_3)_2$	–	(Simonova et al. 2009)
Ordered iron silicate matrix and disordered iron silicate matrix	$CaCl_2$ (7m%)	–	(Ristić and Henninger 2014)
Expanded natural graphite	$CaCl_2$, $SrCl_2$, $BaCl_2$–$FeCl_3$, $NiCl_2$, $MnCl_2$–NH_4Cl, NaBr	–	(Jiang et al. 2013)
Impacted expanded natural graphite ENGB and expanded natural graphite powder ENGP	$CaCl_2$ + KCl	–	(Druske et al. 2014)
Graphite	$SrBr_2$	–	(Lahmidi et al. 2006)
Zeolite 13X	$MgSO_4$	Tests on 200 g of the composite material ZM15 (zeolite 13X impregnated with 15% by mass of $MgSO_4$), charging: 150°C; discharging: 30°C; energy storage density of the composite material: 166 kWh·m^{-3}	(Hongois 2011; Hongois et al. 2011)
Zeolite 13X	$CaCl_2$, $MgSO_4$, $Ca(NO_3)_2$, $Li(NO_3)_2$, LiBr	–	(Casey et al. 2014)

Matrix	Salt	Comments	References
Vermiculite	$CaCl_2$, $MgSO_4$, $Ca(NO_3)_2$, $Li(NO_3)_2$, LiBr	–	(Casey et al. 2014)
Vermiculite and zeolite	$CaCl_2$	–	(Aydin et al. 2016)
Vermiculite	$CaCl_2$	–	(Aristov et al. 2000)
Vermiculite	$LiNO_3$	–	(Sapienza et al. 2012)
MCM-41	$CaCl_2$	–	(Tokarev et al. 2002)
MCM-41 and SBA-15	3, 5 and 7% of Al_3SO_4	–	(Jabbari-Hichri et al. 2016)
Vermiculite	$SrBr_2$	–	(Zhang et al. 2016)
Wakkanai siliceous shale (WSS)	$CaCl_2$	–	(Liu et al. 2013)
WSS	LiCl	–	(Liu et al. 2015)

Table 1.12. *Various composite materials studied for thermochemical heat storage (N'Tsoukpoe 2012b; Marias 2015; Jarimi et al. 2019)*

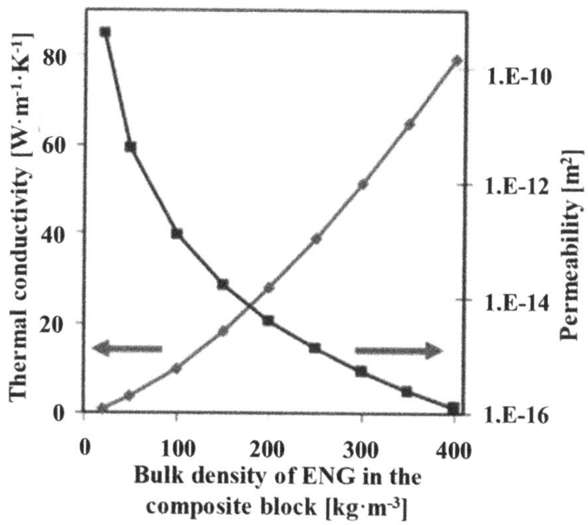

Figure 1.21. *Evolution of effective thermal conductivity and permeability of a reactive composite block as a function of the bulk density of ENG (Stitou 2013). For a color version of this figure, see www.iste.co.uk/lepierres/heat2.zip*

1.7. Overview of material characterization for thermochemical heat storage applications

This section aims to provide an overview of the main material characterization techniques used by research teams focusing on thermochemical heat storage applications. Techniques specifically related to porous materials (porosity measurement, particle size distribution, etc.) are not extensively discussed here. An overview is provided in Table 1.13. Readers interested in these aspects may find useful information in the IUPAC document on key methods for characterizing porous materials (Rouquerol et al. 1994).

Thermal scientists typically seek information that helps specify the thermodynamic conditions for using the material and estimate the energy storage density. This primarily involves determining the thermodynamic equilibrium conditions of the material and estimating the enthalpies of the phenomena.

1.7.1. *Determination of thermodynamic equilibrium conditions*

For solid–gas chemical reactions, one of the initial experimental approaches is the use of TGA to verify if the reaction can occur under the envisaged thermodynamic conditions, such as the sorbate vapor pressure and temperature of the heat source during charging, and the sorbate vapor pressure and temperature during discharging.

TGA also helps determine the stoichiometry of the reaction and its potential steps. It provides an indication of the reversibility of the reaction. TGA is a thermal analysis method that involves tracking the mass evolution of a material sample during heating, cooling, or constant temperature holding under a specified atmosphere. The analyzer includes a highly sensitive balance, a temperature-controlled furnace, a unit for evacuating and controlling the atmosphere, and a data control and recording system. A small amount of material (approximately 10 mg) is deposited in an inert crucible and placed on the analyzer's highly sensitive balance (Figure 1.22).

	Normal
Gas adsorption	Only measures the open pores Pore size: 0.4–50 nm Easy Proven technique
Mercury porosimetry	Similar to the gas adsorption technique Only measures the open pores Pore size > 1.5 nm Easy Proven technique
Transmission electron microscopy (TEM)	Provides information on pore connectivity The size of the pores can be measured if the material contains ordered pores Rarely used for pore analysis
Scanning electron microscopy (SEM)	Pore size > 5 nm Rarely used for pore analysis
Small angle X-ray scattering (SAXS)	All pore sizes Open and closed porosity
Small angle neutron scattering (SANS)	All pore sizes Open and closed porosity Expensive

Table 1.13. *Various techniques for characterizing porous materials*

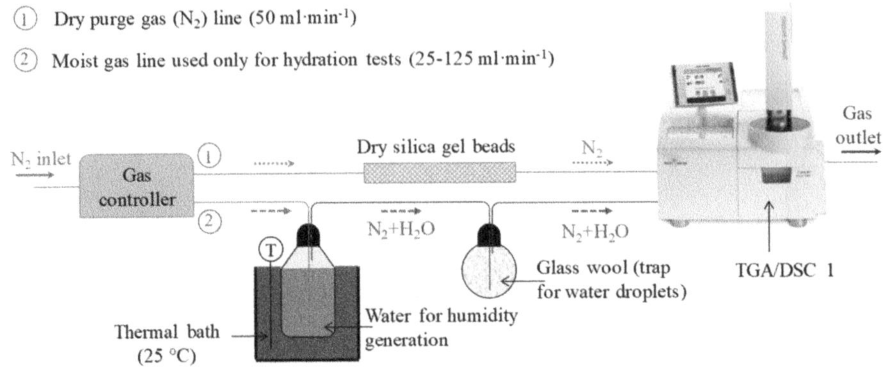

Figure 1.22. *Example of an experimental setup for thermogravimetric analysis (N'Tsoukpoe et al. 2014b). For a color version of this figure, see www.iste.co.uk/lepierres/heat2.zip*

The sample undergoes a temperature variation using a typically inert gas (e.g., nitrogen, helium, or argon) following a predefined program. A temperature probe in direct contact with the crucible allows monitoring the exact temperature of the sample, while its mass is continuously recorded by the electronic balance. Box 1.2 presents an example of TGA results. Interested readers may refer to relevant articles on the subject in the Techniques de l'Ingénieur (Daudon 2001; Wirth et al. 2014).

When it is determined that the material meets a minimum set of criteria, a setup can be arranged to determine a few points on the equilibrium curve of water vapor pressure as a function of temperature.

Figure 1.24 illustrates a device used for characterizing the equilibrium reactions of salt hydrates at low temperatures. In principle, a quantity of salt arranged in a thin layer comes into contact with the vapor phase of the sorbate, which is vaporized at a given temperature using another thermostat bath. The entire system, being previously evacuated by a vacuum pump, has a vapor pressure equal to the saturation pressure of the sorbate at the temperature set by the thermostat bath. This allows for varying the sorbate pressure and measuring the temperature evolution in the salt bed. Of course, it is necessary to allow a "sufficiently long" time for equilibrium to be established, meaning there is no significant temperature variation in the salt bed over a defined period.

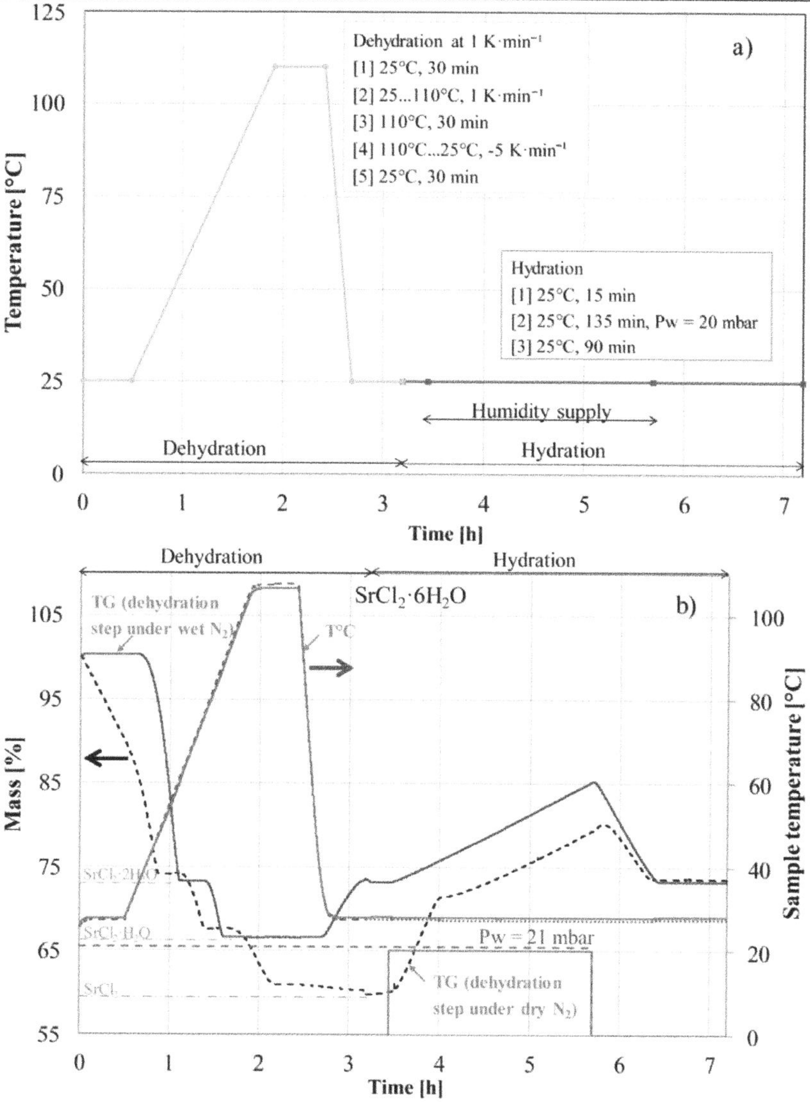

Figure 1.23. *Thermogravimetric analysis of $SrCl_2{\cdot}6H_2O$ (N'Tsoukpoe et al. 2014b). For a color version of this figure, see www.iste.co.uk/lepierres/heat2.zip*

Figure 1.23(b) presents a thermogram of $SrCl_2{\cdot}6H_2O$ subjected to a two-step program (Figure 1.23(a)). In the first step, the material is exposed to a stream of dry nitrogen with a temperature increase from 25°C to approximately 110°C at a heating rate of 1 $K \cdot min^{-1}$. This is followed by a plateau of half an hour at 110°C before the sample is cooled back to 25°C to be exposed to a stream of humid nitrogen with a water vapor pressure of 20 mbar, while maintaining the temperature at 25°C.

The first step allows the observation of the dehydration of $SrCl_2 \cdot 6H_2O$ in three consecutive stages: it first loses 4 moles of water ($SrCl_2 \cdot 6H_2O \rightarrow SrCl_2 \cdot 2H_2O + 4H_2O$), then one mole ($SrCl_2 \cdot 2H_2O \rightarrow SrCl_2 \cdot H_2O + H_2O$) and another mole ($SrCl_2 \cdot H_2O \rightarrow SrCl_2 + H_2O$). The material then begins to react with water when humid nitrogen is introduced into the unit. The process is relatively slow, and when the supply of water vapor is suspended after about 2 h, the salt loses water to stabilize as $SrCl_2 \cdot 2H_2O$.

When the same experiment is conducted by introducing humid nitrogen (water vapor: 20 mbar) into the system from the beginning, dehydration at 110°C leads to the formation of $SrCl_2 \cdot H_2O$, and the final dehydration does not occur.

In conclusion, $SrCl_2 \cdot 6H_2O$ can be completely dehydrated at a temperature below 110°C in the absence of water vapor, for example, with dry air. However, if the operation is carried out under humid air or in a closed system with a condensation pressure of 21 mbar (= condensation temperature of 18°C), the monohydrate ($SrCl_2 \cdot H_2O$) is obtained. Since the experiment was stopped before the end of hydration, no definitive conclusion can be drawn regarding the total reversibility of the reaction under the operating conditions, although the trend of the curve suggests it.

Box 1.2. *Example of thermogravimetric analysis results*

With an aqueous solution, a specially designed device can be used, as presented by Nasirzadeh et al. (2004), allowing the measurement of vapor pressures of aqueous solutions over a wide temperature range (25–200°C) with high pressure precision (±0.3 Pa). Alternatively, vapor pressure analyzers, such as ERAVAP, can be used, as employed by Lefebvre (2015) (Figure 1.25).

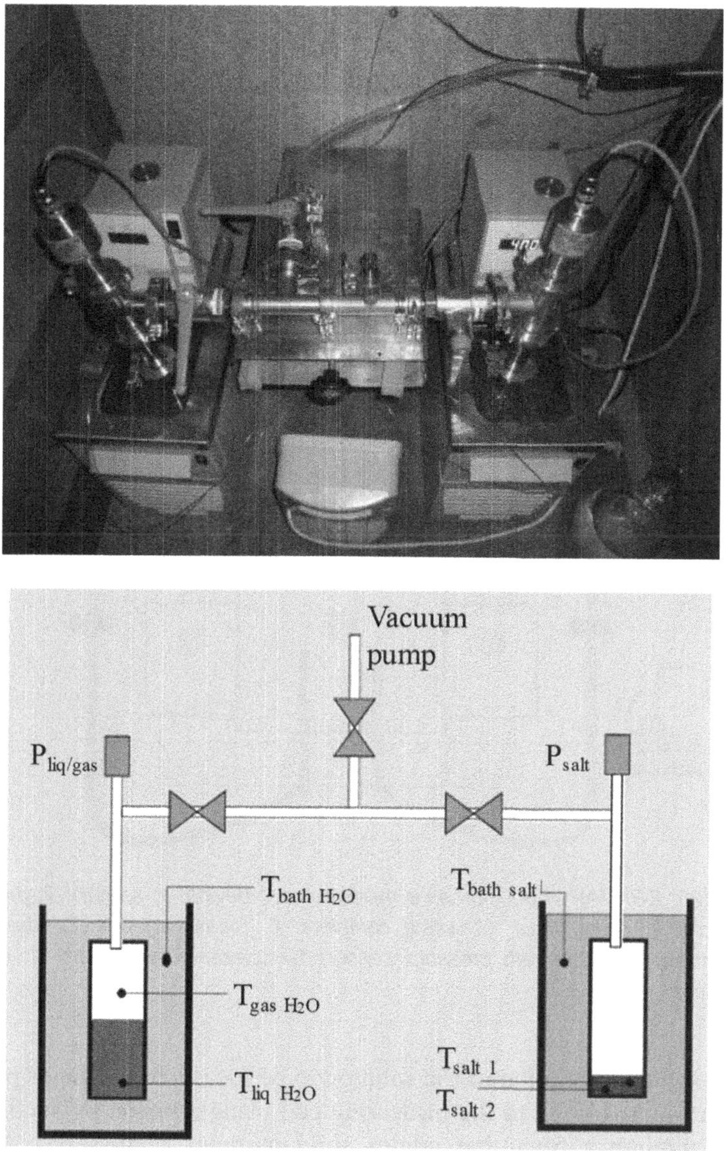

Figure 1.24. *Experimental setup for the characterization of low-temperature reaction equilibria of hydrated salts (Stitou 2013). For a color version of this figure, see www.iste.co.uk/lepierres/heat2.zip*

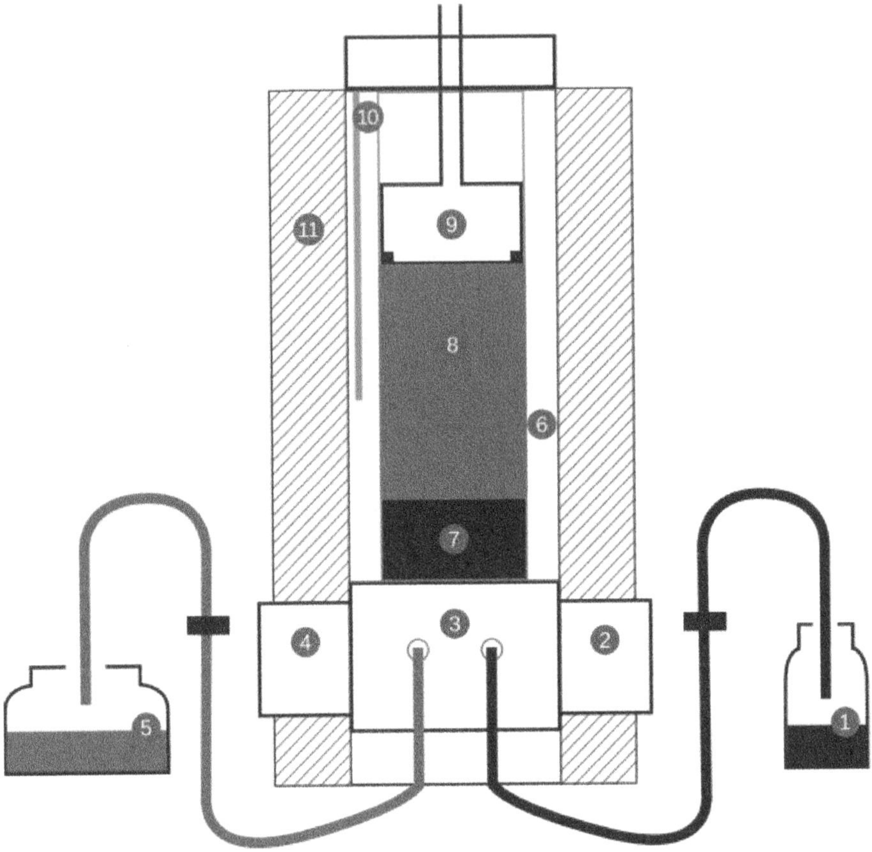

Figure 1.25. *ERAVAP vapor pressure measurement device. 1: sample; 2: inlet valve; 3: manifold; 4: outlet valve; 5: waste container; 6: measuring cell; 7: liquid phase; 8: gas phase; 9: piston with pressure sensor; 10: temperature sensor; 11: insulation (eralytics n.d.)*

For each concentration of the solution to be measured for vapor pressure, a sample is transferred to the measuring cell using a piston by suction after passing through a filter that retains solid particles (undissolved crystals, insoluble particles, etc.). The solution is then brought to the desired measurement temperature. The pressure above the solution is measured by a pressure sensor, and the stabilized value is recorded. Measuring the pressure at three vapor/liquid ratios allows the calculation of the partial pressure of gases dissolved in the solution (especially air) and the deduction of the vapor pressure of the solution. The temperature of the solution is then raised to obtain another measurement point. When finished with a given concentration, the cells are washed multiple times, including with the new solution of different concentration to be measured.

The results obtained by this method by Lefebvre et al. (2014) are shown in Figure 1.26.

For solid–gas physisorbents, there are various methods for determining adsorption isotherms, with the most common being volumetric and gravimetric methods. Other methods include chromatographic techniques, frequency response techniques, closed-loop recycle, column dynamic and isotope exchange (Auerbach et al. 2003; Sircar and Myers 2003). Gravimetric methods appear more realistic and reliable but are relatively slow compared to volumetric methods.

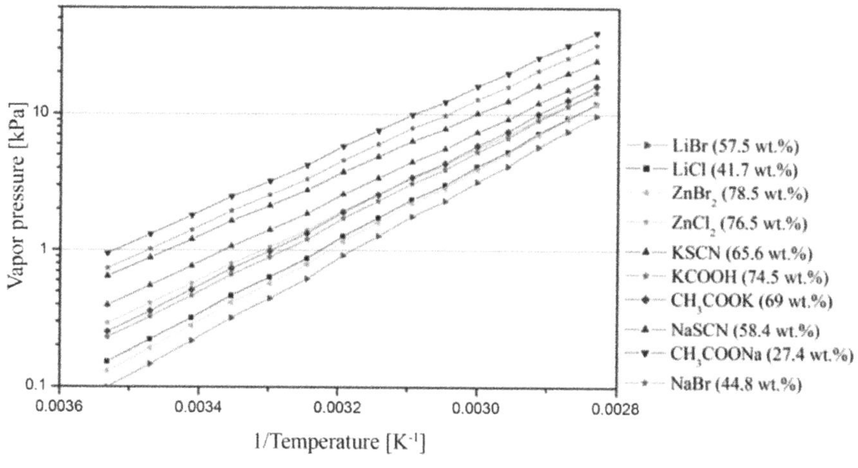

Figure 1.26. *Vapor pressure as a function of the reciprocal of the absolute temperature of aqueous solutions at various mass fractions (Lefebvre et al. 2014). For a color version of this figure, see www.iste.co.uk/lepierres/heat2.zip*

Figure 1.27 presents the principle of these two families of methods. For volumetric methods, the gas pressure is measured before and after expansion of the gas between volumes V_1 and V_2. For the gravimetric method, the mass of the sample is measured with a magnetic suspension balance when raising the pressure. The second orange-coloured mass in the diagram is used to determine both the gas density and to account for buoyancy. The volume of the sample, in red, must be known in both cases. For more details, one can refer to Appendix 1 of Stadie's (2013) doctoral thesis. These experimental data are then correlated by mathematical models that should best represent the experimental adsorption isotherms (Sun et al. 2016). It is desirable for the models to have both physical significance, consistency with fundamental thermodynamic laws, and numerical processing convenience, which can be very useful when simulating adsorption processes.

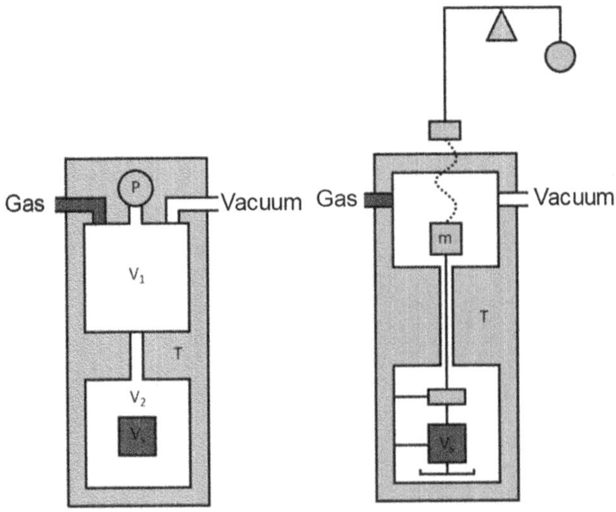

a) b)

Figure 1.27. *Schematic diagram of the apparatus for determining sorption isotherms using the (a) volumetric and (b) gravimetric methods (Stadie 2013). For a color version of this figure, see www.iste.co.uk/lepierres/heat2.zip*

1.7.2. *Determination of enthalpies*

The determination of enthalpies involves calorimetry, and calorimeters are the main equipment to be used here. For chemical reactions, especially those involving salt hydrates or hydroxides, DSC is the primary technique used. It is generally coupled with TGA, and many microcalorimeters offer this dual functionality. A crucible containing the sample is subjected to a temperature program simultaneously with an empty crucible, called a reference, in the same chamber. The chamber is therefore heated or cooled at a constant rate, which varies the temperatures of the sample and the reference. The heat flow at the sample is then proportional to the temperature difference between the sample and the ambient temperature (inside the chamber). The same applies to the reference. For more details, see Mehling and Cabeza (2008) or Grenet and Legendre (2010). The analysis of the thermal flux exchanged by the sample allows the detection of any thermal event that occurs during heating/cooling, as well as the evaluation of the amount of heat involved in this event. An example of a result from DSC is provided in Box 1.3.

Figure 1.28 presents the results obtained from a TGA coupled with a DSC. It involves an initial sample of 11.11 mg of a highly hydrated salt subjected to a five-step program:

1) holding temperature at 25°C for 30 min;

2) heating at a constant rate of 1 K·min^{-1} from 25 to 75°C;

3) holding temperature at 75°C for 60 min;

4) cooling at a constant rate of 10 K·min^{-1} from 75 to 55°C;

5) holding at 25°C for 30 min.

The constant-rate heating starts at t = 30 min and ends at t = 80 min, while the phase where the temperature is held constant ends at t = 140 min. It can be observed that the material loses mass from the beginning of heating, accompanied by heat absorption reaching its peak at around the 74th min. The chemical reaction seems to be completed around t = 110 min and corresponds to a mass loss estimated at 4.0447 mg. The integration of the heat flux corresponding to this mass loss is 12.6 J, giving the enthalpy associated with the observed reaction, that is, 1134 J·g^{-1}. It should be noted that this is a quick assessment, and repetition of experiments is often useful for a good estimation of the quantities involved (see Figure 1.28).

Box 1.3. *Example of a differential scanning calorimetry result*

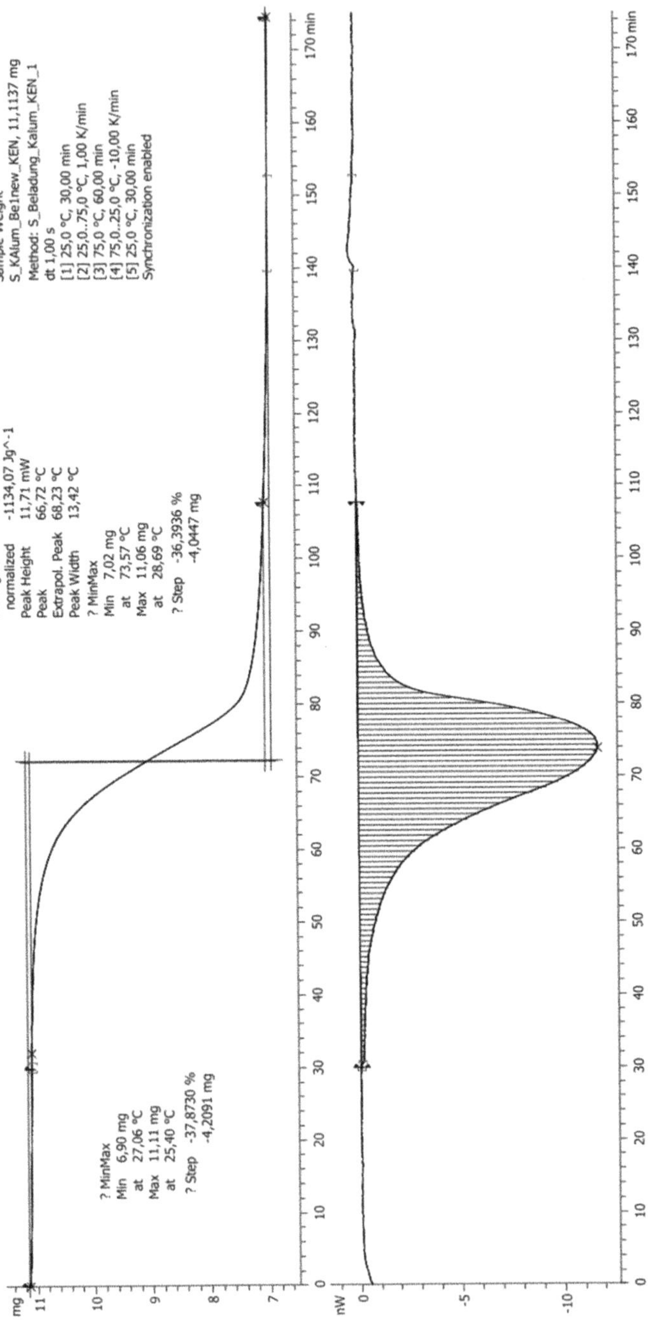

Figure 1.28. *Results of a TGA/DSC of a high salt hydrate*

1.8. References

Adegoke, C.O. (1993). Solubility of the water-lithium-bromide-zinc-bromide combination. *International Journal of Refrigeration*, 16(1), 45–48.

Adokou, K.R.E. (2020). Mise en place d'une bibliothèque de propriétés de matériaux de sorption et identification de matériaux de sorption localement disponibles. Master's Thesis, Institut International d'Ingénierie de l'Eau et de l'Environnement, Ouagadougou.

Alam, K.C.A., Saha, B.B., Akisawa, A., Kashiwagi, T. (2004). Influence of design and operating conditions on the system performance of a two-stage adsorption chiller. *Chemical Engineering Communications*, 191(7), 981–997.

Anies, G. (2011). Modélisation, simulation dynamique, validation expérimentale et optimisation énergétique d'une unité de rafraîchissement solaire par absorption. PhD Thesis, Université de Pau et des Pays de l'Adour, Pau.

Aristov, Y.I. (2007). Novel materials for adsorptive heat pumping and storage: Screening and nanotailoring of sorption properties. *Journal of Chemical Engineering of Japan*, 40(13), 1242–1251.

Aristov, Y.I. and Gordeeva, L.G. (2009). "Salt in a porous matrix" adsorbents: Design of the phase composition and sorption properties. *Kinetics and Catalysis*, 50(1), 65–72.

Aristov, Y.I., Tokarev, M.M., Cacciola, G., Restuccia, G. (1996). Selective water sorbents for multiple applications, 1. $CaCl_2$ confined in mesopores of silica gel: Sorption properties. *Reaction Kinetics and Catalysis Letters*, 59(2), 325–333.

Aristov, Y.I., Restuccia, G., Tokarev, M.M., Buerger, H.D., Freni, A. (2000). Selective water sorbents for multiple applications. 11. $CaCl_2$ confined to expanded vermiculite. *Reaction Kinetics and Catalysis Letters*, 71(2), 377–384.

Auerbach, S.M., Carrado, K.A., Dutta, P.K. (2003). *Handbook of Zeolite Science and Technology*. CRC Press, Boca Raton.

Aydin, D., Casey, S.P., Chen, X., Riffat, S. (2016). Novel "open-sorption pipe" reactor for solar thermal energy storage. *Energy Conversion and Management*, 121, 321–334.

Baerlocher, C., McCusker, L.B., Allen, D.H., Structure Commission of the International Zeolite Association. (2007). *Atlas of Zeolite Framework Types*. Elsevier, Amsterdam [Online]. Available at: www.iza-structure.org/books/Atlas_6ed.pdf [Accessed 2 September 2020].

Bales, C. (2005). Chemical and sorption storage. Selection of concepts. Report B1 of Subtask B, IEA Solar Heating and Cooling Programme [Online]. Available at: http://archive.iea-shc.org/publications/downloads/task32-Chemical_and_Sorption_Storage.pdf [Accessed 4 April 2014].

Bales, C. (2008). Final report of Subtask B. "Chemical and sorption storage". The overview. Report, IEA Solar Heating and Cooling Programme [Online]. Available at: www.iea-shc.org.

Bales, C., Gantenbein, P., Hauer, A., Henning, H.M., Jaenig, D., Kerskes, H., Núñez, T., Visscher, K. (2005). Thermal properties of materials for thermo-chemical storage of solar heat. Report B2 of Subtask B, IEA Solar Heating and Cooling Programme, Borlänge [Online]. Available at: www.iea-shc.org/publications/downloads/task32-Thermal_Properties_of_Materials.pdf.

Bales, C., Gantenbein, P., Jähnig, D., Kerskes, H., Summer, K., van Essen, M., Weber, R. (2008). Laboratory tests of chemical reactions and prototype sorption storage units. Report B4 of Subtask B, IEA Solar Heating and Cooling Programme [Online]. Available at: www.iea-shc.org.

Batten, S.R., Champness, N.R., Chen, X.M., Garcia-Martinez, J., Kitagawa, S., Öhrström, L., O'Keeffe, M., Paik Suh, M., Reedijk, J. (2013). Terminology of metal-organic frameworks and coordination polymers (IUPAC Recommendations 2013). *Pure and Applied Chemistry*, 85(8), 1715–1724.

de Boer, R., Haije, W.G., Veldhuis, J.B.J. (2002). Determination of structural, thermo-dynamic and phase properties in the Na_2S–H_2O system for application in a chemical heat pump. *Thermochimica Acta*, 395(1/2), 3–19.

de Boer, R., Smeding, S.F., Zondag, H.A., Krol, G. (2014). Development of a prototype system for seasonal solar heat storage using an open sorption process. In *Eurotherm Seminar 99: Advances in Thermal Energy Storage*, Lérida [Online]. Available at: www.ecn.nl/publicaties/PdfFetch.aspx?nr=ECN-M–14-009.

Casey, S.P., Elvins, J., Riffat, S., Robinson, A. (2014). Salt impregnated desiccant matrices for 'open' thermochemical energy storage. Selection, synthesis and characterisation of candidate materials. *Energy and Buildings*, 84, 412–425.

Chan, C.W., Ling-Chin, J., Roskilly, A.P. (2013). A review of chemical heat pumps, thermodynamic cycles and thermal energy storage technologies for low grade heat utilisation. *Applied Thermal Engineering*, 50(1), 1257–1273.

Chou, I.M. and Seal, R.R. (2003). Determination of epsomite-hexahydrite equilibria by the humidity-buffer technique at 0.1 MPa with implications for phase equilibria in the system $MgSO_4$-H_2O. *Astrobiology*, 3(3), 619–630.

Darchy, B., Lacour, S., Le Mière, E., Bavoux, E., Domart, Y. (1997). Inhalation de vapeurs d'ammoniaque: valeur pronostique des lésions cutanées. *Réanimation Urgences*, 6(4), 465–469.

Daudon, J.L. (2001). Thermogravimétrie. *Techniques de l'Ingénieur*, P1260 v2, 1–21. doi: 10.51257/a-v2-p1260.

Dicaire, D. and Tezel, F.H. (2009). Study of adsorbent energy density and regeneration for long term thermal energy storage of solar and waste heat. In *Proceedings of the 11th International Conference on Thermal Energy Storage (Effstock 2009)*, Stockholm.

Dicaire, D. and Tezel, F.H. (2011). Regeneration and efficiency characterization of hybrid adsorbent for thermal energy storage of excess and solar heat. *Renewable Energy*, 36, 986–992.

Dirksen, J.A., Ring, T.A., Duvall, K.N., Jongen, N. (2001). Testing of crystallization inhibitors in industrial LiBr solutions. *International Journal of Refrigeration*, 24, 856–859.

Donkers, P.A.J., Sögütoglu, L.C., Huinink, H.P., Fischer, H.R., Adan, O.C.G. (2017). A review of salt hydrates for seasonal heat storage in domestic applications. *Applied Energy*, 199, 45–68.

Druske, M.M. (2020). The reaction of mixed salts in advanced heat storage systems. PhD Thesis, Université Leuphana, Lüneburg [Online]. Available at: https://pubdata.leuphana.de/frontdoor/deliver/index/docId/999/file/The_Reaction_of_Mixed_Salts_in_Advanced_Heat_Storage_Systems.pdf [Accessed 2 September 2020].

Druske, M.M., Fopah-Lele, A., Korhammer, K., Rammelberg, H.U., Wegscheider, N., Ruck, W., Schmidt, T. (2014). Developed materials for thermal energy storage: Synthesis and characterization. *Energy Procedia*, 61, 96–99.

Dyer, A. (2001). Zeolites. In *Encyclopedia of Materials: Science and Technology*, Buschow, K.H.J., Cahn, R.W., Flemings, M.C., Ilschner, B., Kramer, E.J., Mahajan, S., Veyssière, P. (eds). Elsevier, Amsterdam [Online]. Available at: http://www.sciencedirect.com/science/article/pii/B0080431526017848.

Eralytics Eravap (2020). L'analyseur de pression vapeur [Online]. Available at: eralytics.com/wp-content/uploads/1612_eravap_bro_FRANCAIS_web.pdf [Accessed 17 June 2020].

Ferchaud, C.J. (2016). Experimental study of salt hydrates for thermochemical seasonal heat storage. PhD Thesis, Eindhoven University of Technology [Online]. Available at: https://research.tue.nl/files/20452083/20160420_Ferchaud.pdf.

Fernández, R., Ortiz, C., Chacartegui, R., Valverde, J.M., Becerra, J.A. (2019). Dispatchability of solar photovoltaics from thermochemical energy storage. *Energy Conversion and Management*, 191, 237–246.

Fischer, H.R. (2009). What is the best possible heat storage density for a seasonal adsorptive thermal energy storage. In *Proceedings of the 11th International Conference on Thermal Energy Storage (Effstock 2009)*, Stockholm [Online]. Available at: intraweb.stockton.edu/eyos/energy_studies/content/docs/effstock09/posters/163.pdf.

Fumey, B., Weber, R., Gantenbein, P., Daguenet-Frick, X., Williamson, T., Dorer, V. (2014). Closed sorption heat storage based on aqueous sodium hydroxide. *Energy Procedia*, 48, 337–346.

Gaeini, M., Javed, M.R., Ouwerkerk, H., Zondag, H.A., Rindt, C.C.M. (2017). Realization of a 4kW thermochemical segmented reactor in household scale for seasonal heat storage. *Energy Procedia*, 135, 105–114.

Galwey, A.K. and Brown, M.E. (1999). *Thermal Decomposition of Ionic Solids: Chemical Properties and Reactivities of Ionic Crystalline Phases*. Elsevier Science, Amsterdam.

Gantenbein, P. (2008). Fundamental geometrical system structure limitations in a closed adsorption heat storage system. In *Proceedings of the 1st International Conference on Solar Heating, Cooling and Buildings (Eurosun 2008)*, Lisbon.

Gantenbein, P., Brunold, S., Flückiger, F., Frei, U. (2022). Sorption materials for application in solar heat energy storage. Report, Institute für Solartechnik SPF, Université technique de Rapperswil, Rapperswil [Online]. Available at: www.solarenergy.ch.

Gartler, G., Jähnig, D., Purkarthofer, G., Wagner, W. (2004). Development of a high energy density sorption storage system. *Proceedings of Eurosun 2004* [Online]. Available at: aee-intec.at.

Gordeeva, L.G., Tokarev, M.M., Parmon, V.N., Aristov, Y.I. (1998). Selective water sorbents for multiple application, 6. Freshwater production from the atmosphere. *Reaction Kinetics and Catalysis Letters*, 65(1), 153–159.

Gordeeva, L.G., Restuccia, G., Freni, A., Aristov, Y.I. (2002). Water sorption on composites "LiBr in a porous carbon". *Fuel Processing Technology*, 79(3), 225–231.

Gordeeva, L.G., Freni, A., Aristov, Y.I., Restuccia, G. (2009). Composite sorbent of methanol "lithium chloride in mesoporous silica gel" for adsorption cooling machines: Performance and stability evaluation. *Industrial & Engineering Chemistry Research*, 48(13), 6197–6202.

Gordeeva, L.G., Grekova, A., Krieger, T., Aristov, Y.I. (2013). Composites "binary salts in porous matrix" for adsorption heat transformation. *Applied Thermal Engineering*, 50(2), 1633–1638.

Grenet, J. and Legendre, B. (2010). Analyse calorimétrique différentielle à balayage (DSC). *Techniques de l'Ingénieur*, P1205 v1 [Online]. Available at: https://www.techniques-ingenieur.fr/base-documentaire/mesures-analyses-th1/methodes-thermiques-d-analyse-42384210/analyse-calorimetrique-differentielle-a-balayage-dsc-p1205/.

Grevel, K.D. and Majzlan, J. (2009). Internally consistent thermodynamic data for magnesium sulfate hydrates. *Geochimica et Cosmochimica Acta*, 73(22), 6805–6815.

Hadorn, J.C., Bales, C., Streicher, W., Drueck, H. (2007). Projet 32: Concepts de stockage innovants pour bâtiments solaires et basse énergie. *IEA SHC e-Newsletter*, 4.

Hafida, E.H. (2008). Élaboration et caractérisation de solides microporeux de type MeAPO4-5 et MeAPO4-11 (Me: Fe, Co et Sn). PhD Thesis, Université Abdelhamid Ibn Badis Mostaganem, Mostaganem [Online]. Available at: http://e-biblio.univ-mosta.dz/handle/123456789/955 [Accessed 8 December 2020].

Hauer, A. (2002a). Assessment of solid adsorbents in open sorption systems for energetic applications. PhD Thesis, Technical University of Berlin [Online]. Available at: https://depositonce.tu-berlin.de//handle/11303/766 [Accessed 3 August 2018].

Hauer, A. (2002b). Thermal energy storage with zeolite for heating and cooling applications. In *3rd Workshop of Annex 17 ECES IA/IEA, Advanced Thermal Energy Storage through Phase Change Materials and Chemical Reactions – Feasibility Studies and Demonstration Projects*, Tokyo.

Hauer, A. (2007a). Adsorption systems for TES-Design and demonstration projects. In *NATO Science Series*, Paksoy, H.Ö. (ed.). Springer, Dordrecht [Online]. Available at: www.springerlink.com/content/r78t5x6kn885735t/.

Hauer, A. (ed.) (2007b). Sorption theory for thermal energy storage. In *Thermal Energy Storage for Sustainable Energy Consumption*, Paksoy, H.O. (ed.). Springer, Dordrecht, 393–408.

Hauer, A. (2010). Compact thermal storages: Potentials and limitations for different applications. In *Proceedings of the 2nd International Conference on Solar Heating, Cooling and Buildings (Eurosun 2010)*, San Francisco.

van Helden, W. and Rommel, M. (2015). Compact thermal energy storage: Material development for system integration. IEA SHC Position Paper, Task 42 (ECES Annex 29) [Online]. Available at: www.iea-shc.org/Data/Sites/1/publications/IEA-SHC-Compact-Thermal-Storage-Position-Paper.pdf [Accessed 30 December 2020].

Hiu, L., N'Tsoukpoe, K.E., Le Pierres, N., Luo, L. (2011). Evaluation of a seasonal storage system of solar energy for house heating using different absorption couples. *Energy Conversion and Management*, 52(6), 2427–2436.

Hiu, L., Nagano, K., Sugiyama, D., Togawa, J., Nakamura, M. (2013). Honeycomb filters made from mesoporous composite material for an open sorption thermal energy storage system to store low-temperature industrial waste heat. *International Journal of Heat and Mass Transfer*, 65, 471–480.

Hiu, L., Nagano, K., Togawa, J. (2015). A composite material made of mesoporous siliceous shale impregnated with lithium chloride for an open sorption thermal energy storage system. *Solar Energy*, 186–200.

Hongois, S. (2011). Stockage de chaleur inter-saisonnier par voie thermochimique pour le chauffage solaire de la maison individuelle. PhD Thesis, Institut National des Sciences Appliquées, Lyon.

Hongois, S., Kuznik, F., Stevens, P., Roux, J.J. (2011). Development and characterisation of a new $MgSO_4$-zeolite composite for long-term thermal energy storage. *Solar Energy Materials and Solar Cells*, 95(7), 1831–1837.

Humphries, T.D., Møller, K.T., Rickard, W.D.A., Sofianos, M.V., Liu, S., Buckley, C.E., Paskevicius, M. (2019). Dolomite: A low cost thermochemical energy storage material. *Journal of Materials Chemistry A*, 7(3), 1206–1215.

International Union of Pure and Applied Chemistry (2014). Compendium of chemical terminology: Gold book. Report, International Union of Pure and Applied Chemistry, Oxford.

Jabbari-Hichri, A., Bennici, S., Auroux, A. (2015). Enhancing the heat storage density of silica–alumina by addition of hygroscopic salts ($CaCl_2$, $Ba(OH)_2$, and $LiNO_3$). *Solar Energy Materials and Solar Cells*, 140, 351–360.

Jabbari-Hichri, A., Bennici, S., Auroux, A. (2016). Effect of aluminum sulfate addition on the thermal storage performance of mesoporous SBA-15 and MCM-41 materials. *Solar Energy Materials and Solar Cells*, 149, 232–241.

Jähnig, D., Hausner, R., Wagner, W., Isaksson, C. (2006). Thermo-chemical storage for solar space heating in single-family house. In *Proceedings of the 10th International Conference on Thermal Energy Storage, Ecostock 2006*, New Jersey.

Jänchen, J., Ackermann, D., Stach, H., Brösicke, W. (2004). Studies of the water adsorption on zeolites and modified mesoporous materials for seasonal storage of solar heat. *Solar Energy*, 76(1/3), 339–344.

Jänchen, J., Ackermann, D., Weiler, E., Stach, H., Brösicke, W. (2005). Calorimetric investigation on zeolites, $AlPO_4$'s and $CaCl_2$ impregnated attapulgite for thermochemical storage of heat. *Thermochimica Acta*, 434(1/2), 37–41.

Jarimi, H., Aydin, D., Yanan, Z., Ozankaya, G., Chen, X., Riffat, S. (2019). Review on the recent progress of thermochemical materials and processes for solar thermal energy storage and industrial waste heat recovery. *International Journal of Low-Carbon Technologies*, 14(1), 44–69.

Jiang, L., Wang, L.W., Jin, Z.Q., Wang, R.Z., Dai, Y.J. (2013). Effective thermal conductivity and permeability of compact compound ammoniated salts in the adsorption/desorption process. *International Journal of Thermal Sciences*, 71, 103–110.

de Jong, A.J., Trausel, F., Finck, C., van Vliet, L., Cuypers, R. (2014). Thermochemical heat storage – System design issues. *Energy Procedia*, 48, 309–319.

Kerskes, H. (2006). Seasonal sorption heat storage. In *Danvak seminar (Solar heating systems - Combisystems - Heat storage)*, DTU Lyngby.

Kerskes, H., Asenbeck, S., Mette, B., Bertsch, F., Müller-Steinhagen, H. (2010). Experimental and numerical investigations on thermo-chemical heat storage. In *Proceedings of the 2nd International Conference on Solar Heating, Cooling and Buildings (Eurosun 2010)*, Graz [Online]. Available at: task42.iea-shc.org/data/sites/1/publications/Task42-Experimental_and_Numerical_Investi-gations_on_Thermo-Chemical_Heat_Storage.pdf [Accessed 14 September 2014].

Krese, G., Kozelj, R., Butala, V., Stritih, U. (2018). Thermochemical seasonal solar energy storage for heating and cooling of buildings. *Energy & Buildings*, 164, 239–253.

Lahmidi, H., Mauran, S., Goetz, V. (2006). Definition, test and simulation of a thermochemical storage process adapted to solar thermal systems. *Solar Energy*, 80(7), 883–893.

Lass-Seyoum, A., Borozdenko, D., Friedrich, T., Langhof, T., Mack, S. (2016). Practical test on a closed sorption thermochemical storage system with solar thermal energy. *Energy Procedia*, 91, 182–189.

Le Pierrès, N. (2014). Procédés à sorption pour la conversion, le stockage et le transport de l'énergie thermique : du matériau à l'intégration énergétique du procédé. HDR, Université de Grenoble, Grenoble.

Lee, H.R., Koo, K.K., Jeong, S., Kim, J.S., Lee, H., Oh, Y.S., Park, D.R., Baek, Y.S. (2000). Thermodynamic design data and performance evaluation of the water + lithium bromide + lithium iodide + lithium nitrate + lithium chloride system for absorption chiller. *Applied Thermal Engineering*, 20(8), 707–720.

Lefebvre, E. (2015). Procédé par absorption avec stockage d'énergie solaire intersaisonnier intensifié par la cristallisation de l'absorbant : recherche & caractérisation thermodynamique de nouveaux couples; conception de la cuve de stockage. PhD Thesis, Université de Lyon 1, Lyon.

Lefebvre, E., Gagnière, E., Bennici, S., Auroux, A., Mangin, D. (2014). Étude des propriétés thermodynamiques de solutions salines dans le cadre d'un procédé de stockage inter-saisonnier de l'énergie solaire. In *Actes du Congrès de la Société Française de Thermique*, Lyon [Online]. Available at: www.sft.asso.fr/Local/sft/dir/user-3775/documents/actes/Congres_2014/Communi-cations/8836.pdf [Accessed 24 August 2018].

Leppäjärvi, T., Malinen, I., Kangas, J., Tanskanen, J. (2012). Utilization of P_i^{sat} temperature-dependency in modelling adsorption on zeolites. *Chemical Engineering Science*, 69(1), 503–513.

Liu, D., Xin-Feng, L., Bo, L., Si-quan, Z., Yan, X. (2018). Progress in thermochemical energy storage for concentrated solar power: A review. *International Journal of Energy Research*, 42(15), 4546–4561.

Marias, F.E. (2015). Analyse, conception et expérimentation de procédés de stockage thermique résidentiel de longue durée par réaction thermochimique à pression atmosphérique. PhD Thesis, Université Grenoble Alpes, Grenoble [Online]. Available at: https://tel.archives-ouvertes.fr/tel-01207191/document [Accessed 5 September 2018].

Marias, F.E., Neveu, P., Tanguy, G., Papillon, P. (2014). Thermodynamic analysis and experimental study of solid/gas reactor operating in open mode. *Energy*, 66, 757–765.

Mauran, S., Lahmidi, H., Goetz, V. (2008). Solar heating and cooling by a thermochemical process. First experiments of a prototype storing 60 kWh by a solid/gas reaction. *Solar Energy*, 82(7), 623–636.

Mehling, H. and Cabeza, L.F. (2008). *Heat and Cold Storage with PCM. An Up To Date Introduction into Basics and Applications*. Springer-Verlag, Berlin.

Metchueng Kamdem, S. (2016). Stockage de chaleur dans l'habitat par sorption zéolithe/H_2O. PhD Thesis, Université de Lyon, Lyon [Online]. Available at: theses.insa-lyon.fr/publication/2016LYSEI059/these.pdf [Accessed 8 May 2018].

Mette, B., Kerskes, H., Drück, H., Badenhop, T., Salg, F., Gläser, R. (2013). Thermochemical energy storage as an element for the energy turnaround. In *Proceedings of the 8th International Renewable Energy Storage Conference and Exhibition (IRES 2013)*, Berlin.

Michel, B. (2012). Procédé thermochimique pour le stockage intersaisonnier de l'énergie solaire : modélisation multi-échelles et expérimentation d'un prototype sous air humide. PhD Thesis, Université de Perpignan Via Domitia, Perpignan.

Michel, B., Mazet, N., Mauran, S., Stitou, D., Xu, J. (2012). Thermochemical process for seasonal storage of solar energy: Characterization and modeling of a high density reactive bed. *Energy*, 47(1), 553–563.

Michel, B., Mazet, N., Neveu, P. (2014). Experimental investigation of an innovative thermochemical process operating with a hydrate salt and moist air for thermal storage of solar energy: Global performance. *Applied Energy*, 129, 177–186.

Mugnier, D. (2013). Market available chillers for solar cooling. *Solar Update*, 57, 12.

Nasirzadeh, K., Zimin, D., Neueder, R., Kunz, W. (2004). Vapor-pressure measurements of liquid solutions at different temperatures: Apparatus for use over an extended temperature range and some new data. *Journal of Chemical & Engineering Data*, 49(3), 607–612.

Nicolas, S., Lopez, K., Lutz, C., Bouvier, L. (2015). Zéolithes à porosité hiérarchisée. Patent, WO2015019013A2 [Online]. Available at: https://patents.google.com/patent/WO2015019013A2/fr.

Nonnen, T., Beckert, S., Gleichmann, K., Brandt, A., Unger, B., Kerskes, H., Mette, B., Bonk, S., Badenhop, T., Salg, F., Gläser, R. (2016). Expérimentation d'un stockage thermochimique à long terme à base d'un composite zéolithe-sel. *Chemie Ingenieur Technik*, 88(3), 363–371.

N'Tsoukpoe, K.E. (2012). Étude du stockage à long terme de l'énergie solaire thermique par procédé d'absorption LiBr–H_2O pour le chauffage de l'habitat. PhD Thesis, Université de Grenoble, Le Bourget du Lac [Online]. Available at: https://tel.archives-ouvertes.fr/tel-00710249.

N'Tsoukpoe, K.E. (2018). Énergie solaire thermique : production, utilisation et rationalisation de la gestion. HDR, Université Grenoble Alpes, Grenoble.

N'Tsoukpoe, K.E. and Kuznik, F. (2021). A reality check on long-term thermochemical heat storage for household applications. *Renewable and Sustainable Energy Reviews*, 139, 110683.

N'Tsoukpoe, K.E., Le Pierrès, N., Luo, L. (2012). Numerical dynamic simulation and analysis of a lithium bromide/water long-term solar heat storage system. *Energy*, 37(1), 346–358.

N'Tsoukpoe, K.E., Restuccia, G., Schmidt, T., Py, X. (2014a). The size of sorbents in low pressure sorption or thermochemical energy storage processes. *Energy*, 77, 983–998.

N'Tsoukpoe, K.E., Schmidt, T., Rammelberg, H.U., Watts, B.A., Ruck, W.K.L. (2014b). A systematic multi-step screening of numerous salt hydrates for low temperature thermochemical energy storage. *Applied Energy*, 124, 1–16.

N'Tsoukpoe, K.E., Osterland, T., Opel, O., Ruck, W.K.L. (2016). Cascade thermochemical storage with internal condensation heat recovery for better energy and exergy efficiencies. *Applied Energy*, 181, 562–574.

Office fédéral de l'environnement (2015). Installations de réfrigération: Prudence avec l'ammoniac [Online]. Available at: https://www.bafu.admin.ch/bafu/fr/home/themen/thema-stoerfallvorsorge/stoerfaelle–dossiers/stoerfallvorsorge–kaelteanlagen-mit-ammoniak–sicherheit-ist-ei.html [Accessed 5 May 2020].

Pardo, P., Deydier, A., Anxionnaz-Minvielle, Z., Rougé, S., Cabassud, M., Cognet, P. (2014). A review on high temperature thermochemical heat energy storage. *Renewable and Sustainable Energy Reviews*, 32, 591–610.

Pataskar, S.G., Adyanthaya, S.D., Devotta, S., Holland, F.A. (2011). Performance of an experimental absorption heat transformer using aqueous lithium bromide, lithium chloride, and lithium bromide/lithium chloride solutions. *Industrial & Engineering Chemistry Research*, 29, 1658–1662.

Patek, J. and Klomfar, J. (2006). A computationally effective formulation of the thermodynamic properties of LiBr–H_2O solutions from 273 to 500 K over full composition range. *International Journal of Refrigeration*, 29, 566–578.

Payra, P. and Dutta, P.K. (2003). Zeolites: A primer. In *Handbook of Eolite Science and Technology*, Auerbach, S.M., Carrado, K.A., Dutta, P.K. (eds). CRC Press, Boca Raton.

Posern, K. and Kaps, C. (2010). Calorimetric studies of thermochemical heat storage materials based on mixtures of $MgSO_4$ and $MgCl_2$. *Thermochimica Acta*, 502(1), 73–76.

Rambaud, G. (2009). Problématique des transferts en milieu poreux réactif déformable pour procédés de rafraîchissement solaire. PhD Thesis, Université de Perpignan, Perpignan.

Rao, N.K., Hanson, J., Dulloo, M.E., Ghosh, K., Nowell, A., Larinde, M. (2006). Manual of seed handling in genebanks. Report, Bioversity International, Rome [Online]. Available at: www.bioversityinternational.org/fileadmin/user_upload/online_library/publications/pdfs/1167.pdf.

Rathgeber, C., Lävemann, E., Hauer, A. (2015). Economic top–down evaluation of the costs of energy storages – A simple economic truth in two equations. *Journal of Energy Storage*, 2, 43–46.

Rathgeber, C., Hiebler, S., Lävemann, E., Dolado, P., Lazaro, A., Gasia, J., de Gracia, A., Miró, L., Cabeza, L.F., König-Haagen, A. et al. (2016). IEA SHC Task 42/ECES Annex 29 – A simple tool for the economic evaluation of thermal energy storages. *Energy Procedia*, 91, 197–206.

Richter, M., Habermann, E.M., Siebecke, E., Linder, M. (2018). A systematic screening of salt hydrates as materials for a thermochemical heat transformer. *Thermochimica Acta*, 659, 136–150.

Ring, T.A., Dirksen, J.A., Duvall, K.N., Jongen, N. (2001). $LiBr{\cdot}2H_2O$ crystallization inhibition in the presence of additives. *Journal of Colloid and Interface Science*, 239, 399–408.

Ristić, A. and Henninger, S.K. (eds) (2014). Sorption composite materials for solar thermal. In *Energy Procedia*. Elsevier, Amsterdam.

Rouquerol, J., Avnir, D., Fairbridge, C.W., Everett, D.H., Haynes, J.M., Pernicone, N., Ramsay, J.D.F., Sing, K.S.W., Unger, K.K. (1994). Recommendations for the characterization of porous solids (Technical Report). *Pure and Applied Chemistry*, 66(8), 1739–1758.

Rouquerol, F., Rouquerol, J., Sing, K. (1999). *Adsorption by Powder and Porous Solids: Principles, Methodology and Applications.* Academic Press, London.

Ruthven, D.M. (1984). *Principles of Adsorption and Adsorption Processes.* John Wiley & Sons, New York.

Sapienza, A., Glaznev, I.S., Santamaria, S., Freni, A., Aristov, Y.I. (2012). Adsorption chilling driven by low temperature heat: New adsorbent and cycle optimization. *Applied Thermal Engineering*, 32, 141–146.

Scapino, L., Zondag, H.A., Van Bael, J., Diriken, J., Rindt, C.C.M. (2017). Sorption heat storage for long-term low-temperature applications: A review on the advancements at material and prototype scale. *Applied Energy*, 190, 920–948.

Sharma, B. (2002). *Krishan's Objective Pre Engineering Chemistry*. Krishna Prakashan Media, Meerut.

Silva, V.M.T.M. and Rodrigues, A.E. (1999). Adsorption and diffusion in bidisperse pore structures. *Industrial & Engineering Chemistry Research*, 38(10), 4023–4031.

Simonova, I.A., Freni, A., Restuccia, G., Aristov, Y.I. (2009). Water sorption on composite "silica modified by calcium nitrate". *Microporous and Mesoporous Materials*, 122(1), 223–228.

Sircar, S. (2006). Gas sorption kinetics by differential closed-loop recycle method: Effect of heat of adsorption. *Adsorption*, 12(4), 259–266.

Sircar, S. and Myers, A.L. (2003). Gas separation by zeolites. In *Handbook of Zeolite Science and Technology*, Auerbach, S.M., Carrado, K.A., Dutta, P.K. (eds). CRC Press, Boca Raton.

Sögütoglu, L.C., Donkers, P.A.J., Fischer, H.R., Huinink, H.P., Adan, O.C.G. (2018). In-depth investigation of thermochemical performance in a heat battery: Cyclic analysis of K_2CO_3, $MgCl_2$ and Na_2S. *Applied Energy*, 215, 159–173.

Stadie, N.P. (2013). Synthesis and thermodynamic studies of physisorptive energy storage materials. PhD Thesis, California Institute of Technology, Pasadena [Online]. Available at: https://resolver.caltech.edu/CaltechTHESIS:09092012-010239493 [Accessed 8 December 2020].

Stitou, D. (2013). Transformation, conversion, stockage, transport de l'énergie thermique par procédés thermochimiques et thermo-hydrauliques. HDR, Université de Perpignan Via Domitia, Perpignan [Online]. Available at: https://tel.archives-ouvertes.fr/tel-00841655.

Stutz, B., Le Pierres, N., Kuznik, F., Johannes, K., Palomo Del Barrio, E., Bédécarrats, J.P., Gibout, S., Marty, P., Zalewski, L., Soto, J., et al. (2017). Storage of thermal solar energy. *Comptes Rendus Physique*, 18(7), 401–414.

Sugimoto, K., Dinnebier, R.E., Hanson, J.C. (2007). Structures of three dehydration products of bischofite from in situ synchrotron powder diffraction data ($MgCl_2{\cdot}nH_2O$; n = 1, 2, 4). *Acta Crystallographica Section B Structural Science*, 63(2), 235–242.

Sun, L.M. and Meunier, F. (1987). Non-isothermal adsorption in a bidisperse adsorbent pellet. *Chemical Engineering Science*, 42(12), 2899–2907.

Sun, L.M., Meunier, F. (2003). Adsorption – Aspects théoriques. *Techniques de l'Ingénieur*, J2730, 1–16. doi: 10.51257/a-v1-j2730.

Sun, L.M., Meunier, F., Brodu, N., Manero, M.H. (2016). Adsorption – Aspects théoriques. *Techniques de l'Ingénieur* [Online]. Available at: www.techniques-ingenieur.fr/base-documentaire/procedes-chimie-bio-agro-th2/operations-unitaires-extractions-fluide-fluide-et-fluide-solide-42332210/adsorption-j2730/.

Tanashev, Y.Y., Krainov, A.V., Aristov, Y.I. (2013). Thermal conductivity of composite sorbents "salt in porous matrix" for heat storage and transformation. *Applied Thermal Engineering*, 61(2), 401–407.

Tatsidjodoung, P., Le Pierrès, N., Luo, L., Heintz, J., Lagre, D. (2014). Étude du fonctionnement d'un réacteur de zéolithe pour le stockage thermochimique de chaleur. In *Actes du Congrès de la Société Française de Thermique*. Société Française de Thermique, Lyon.

Teo, H.W.B., Chakraborty, A., Kayal, S. (2017). Post synthetic modification of MIL-101(Cr) for S-shaped isotherms and fast kinetics with water adsorption. *Applied Thermal Engineering*, 120, 453–462.

Thomas, W.J. and Crittenden, B. (1998). *Adsorption Technology and Design*. Elsevier, Amsterdam.

Tian, B., Jin, Z.Q., Wang, L.W., Wang, R.Z. (2012). Permeability and thermal conductivity of compact chemical and physical adsorbents with expanded natural graphite as host matrix. *International Journal of Heat and Mass Transfer*, 55(15/16), 4453–4459.

Tokarev, M., Gordeeva, L., Romannikov, V., Glaznev, I., Aristov, Y.I. (2002). New composite sorbent $CaCl_2$ in mesopores for sorption cooling/heating. *International Journal of Thermal Sciences*, 41(5), 470–474.

Tournoud, C. (2008). Les intoxications accidentelles domestiques par inhalation. In *Actes de URGENCES 2008, 2e Congrès de la Société Française de Médecine d'Urgence (SFMU)*, Paris [Online]. Available at: www.sfmu.org/upload/70_formation/02_eformation/02_congres/Urgences/urgences2008/donnees/pdf/045_tournoud.pdf [Accessed 5 May 2020].

Tso, C.Y. and Chao, C.Y.H. (2012). Activated carbon, silica-gel and calcium chloride composite adsorbents for energy efficient solar adsorption cooling and dehumidification systems. *International Journal of Refrigeration*, 35(6), 1626–1638.

Wagman, D.D., Evans, W.H., Parker, V.B., Schumm, R.H., Halow, I., Bailey, S.M., Churney, K.L., Nuttall, R.L. (1982). The NBS tables of chemical thermodynamic properties. Selected values for inorganic and C1 and C2 organic substances in SI units. *Journal of Physical and Chemical Reference Data*, 11, 2385.

Wang, X., Zimmermann, W., Ng, K.C., Chakraboty, A., Keller, J.U. (2004). Investigation on the isotherm of silica gel + water systems. TG and volumetric methods. *Journal of Thermal Analysis and Calorimetry*, 76, 659–669.

Wang, R., Wang, L., Wu, J. (eds) (2014). Mechanism and thermodynamic properties of physical adsorption. In *Adsorption Refrigeration Technology*. John Wiley & Sons, New York [Online]. Available at: onlinelibrary.wiley.com/doi/abs/10.1002/9781118197448.ch3.

Weber, R. and Dorer, V. (2008). Long-term heat storage with NaOH. *Vacuum*, 82, 708–716.

Weber, R., Asenbeck, S., Kerskes, H., Drück, H. (2016). SolSpaces – Testing and performance analysis of a segmented sorption store for solar thermal space heating. *Energy Procedia*, 91, 250–258.

Whiting, G., Grondin, D., Bennici, S., Auroux, A. (2013). Heats of water sorption studies on zeolite $MgSO_4$ composites as potential thermochemical heat storage materials. *Solar Energy Materials and Solar Cells*, 112, 112–119.

Wirth, E., Guitteny, F., Mathonat, C. (2014). Thermogravimétrie. *Techniques de l'Ingénieur*, 1–21 [Online]. Available at: www.techniques-ingenieur.fr/base-documentaire/mesures-analyses-th1/methodes-thermiques-d-analyse-42384210/thermogravimetrie-p1260/.

Wong, B., van Helden, W., Rosen, M.A. (2013). Compact thermal energy storage for potential Canadian market applications. In *3rd Climate Change Technology Conference, CCTC 2013*, Montreal [Online]. Available at: www.cctc2013.ca/Papers/CCTC2013%20TRA1-3%20Wong.pdf.

W. R. Grace & Co.-Conn. (2010). GRACE Davison/SYLOBEAD. Adsorbents for process applications [Online]. Available at: grace.com/general-industrial/en-us/Documents/sylobead_br_E_2010_f100222_web.pdf [Accessed 1 September 2020].

Wu, H., Wang, S., Zhu, D. (2007). Effects of impregnating variables on dynamic sorption characteristics and storage properties of composite sorbent for solar heat storage. *Solar Energy*, 81(7), 864–871.

Yu, N., Wang, R.Z., Lu, Z.S., Wang, L.W., Ishugah, T.F. (2014). Evaluation of a three-phase sorption cycle for thermal energy storage. *Energy*, 67, 468–478.

Zettl, B., Englmair, G., Steinmaurer, G. (2014). Development of a revolving drum reactor for open-sorption heat storage processes. *Applied Thermal Engineering*, 70(1), 42–49.

Zhang, Y.N., Wang, R.Z., Zhao, Y.J., Li, T.X., Riffat, S.B., Wajid, N.M. (2016). Development and thermochemical characterizations of vermiculite/SrBr2 composite sorbents for low-temperature heat storage. *Energy*, 115, 120–128.

Zheng, X., Ge, T.S., Wang, R.Z. (2014). Recent progress on desiccant materials for solid desiccant cooling systems. *Energy*, 74, 280–294.

Zondag, H.A., van Essen, V.M., Bakker, M., Bach, P.W. (2010). An evaluation of the economical feasibility of seasonal sorption heat storage. In *Proceedings of the 5th International Renewable Energy Storage Conference (IRES 2010)* [Online]. Available at: www.ecn.nl/docs/library/report/2010/m10096.pdf.

2

Heat Storage Using Absorption Processes

Nolwenn Le Pierrès
LOCIE, CNRS, Université Savoie Mont Blanc, Le Bourget-du-Lac, France

From among the thermochemical storage processes, we mainly use fluids (liquids and gases): the absorption process. This storage method is currently only in the development stage and has not yet been implemented commercially. For the time being, the processes for storing heat through absorption have only been developed to store "low temperature" heat. For the purposes of giving consideration to sustainable development, both of the following must be implemented:

– to make better use of the energy used in current processes, and particularly heat energy known as "waste" heat, which is currently discharged into the surrounding environment without being used. The amounts of heat disposed of in this way are considerable. For perspective, the discarded industrial waste heat in France totals 109.5 TWh each year, which is equal to 36% of the heat needed by this same industrial sector (ADEME 2017);

– to develop the use of renewable energies, and in particular those that involve heat, such as solar thermal energy, or geothermal energy.

For a color version of all the figures in this chapter, see www.iste.co.uk/lepierres/heat2.zip.

Heat and Cold Storage 2,
coordinated by Nolwenn Le Pierrès and Lingai Luo.

In both cases, the temperature ranges involved are less than 200°C, and often less than 100°C (ADEME 2017). This temperature range, while not suitable to be used very efficiently for uses such as producing electricity, is nevertheless very useful to the form of heat in many industrial processes, as mentioned above; or in the case of lower temperatures, for heating buildings, whether used directly in the building itself, or via a district heating network. In fact, buildings (residential and commercial) are the largest sector of energy consumption from among all economic sectors combined. They represent 40% of end energy consumption in France, using 720 TWh in 2018 (RTE 2020), and the energy consumption in this sector sees significant seasonal and hourly variations (summer/winter, off-peak hours/working hours). Thus, the field of application for "low temperature" heat storage is very broad, both from the point of view of resources and of needs.

The principle of the absorption heat storage system is based on that of the absorption heat pump (AHP), which for its part is well known. On the basis of this heat pump, adding tanks to the system makes it possible to store heat for short or long periods without any loss of thermochemical potential over time. The energy density of this type of storage is also particularly noteworthy. In this chapter, we will present the principles of these processes, their operation, how their performances are quantified, their critical components (the sorption exchangers), intensified cycles and examples of integration.

2.1. Absorption processes: the principle

The principle of absorption storage processes is based on that of heat pumps.

2.1.1. *AHPs*

AHP cycles have been known and implemented in practical uses since the mid-19th century. They can be used over a wide temperature range, both for the production of cold and heat. Currently, these systems are well developed. They have been and remain the subject of numerous scientific studies and publications (Altamirano et al. 2019; Nikbakhti et al. 2020). In fact, absorption machines currently provide for roughly half of the demand for air conditioning in Japan (Herold et al. 2016). Likewise, in the field of refrigeration, they occupy niche markets, such as refrigerators for trailers, motor homes, or hotel rooms thanks to their ability to operate without electricity, and their reduced noise during operation (which is much quieter than vapor compression refrigerators). Current developments on AHPs seek to expand the range of use of these machines, particularly in machines with lower capacities (on the order of 10 kW) for domestic applications, to increase their compactness, and to use them in conjunction with renewable heat sources such as solar.

In AHP cycles, the sorption that occurs is the linking of a gas (also called a "working fluid" or "sorbate") with a liquid (the "sorbent"). The working fluid and the sorbent form the "sorption couple". These two fluids can be combined; this is known as a physical sorption (or physisorption), involving weak Van der Waals bonds between the sorbent and the sorbate. When these two fluids are combined in liquid form, the sorbent-sorbate mixture is called a solution. The amount of working fluid sorbated in the solution is quantified by the mass fraction of sorbent in the sorbent-working fluid solution. This mass fraction can also be called sorbent concentration. This mass fraction varies between specific limits, which are functions of the sorption pair called for and of the operating conditions of the couple in the machine. The terminology used varies between different authors, but the most frequent convention (which will be adopted here) is to categorize solutions with a high mass fraction of the sorbent as "rich" or "concentrated", and a solution with large amounts of working fluid as "poor" or "diluted", and therefore containing a low mass fraction of sorbent.

The absorption is a reversible physical phenomenon and a thermal phenomenon as well. It can be represented by the equation:

$$\text{Rich solution of } AB_{liquid} + \text{working fluid } B_{steam} \leftrightarrow \text{Poor solution } AB_{liquid} + \text{heat} \quad [2.1]$$

In the direct sense of equation [2.1], the phenomenon of absorption of the vapor B into the solution AB is exothermic. Thus, the mass fraction of sorbent A in the solution decreases. When the phenomenon is reversed (meaning that it "returns" in the direction of equation [2.1]), a heat supply allows for an endothermal desorption of B from the solution AB, or to put it differently, the separation of B in the gaseous form from the solution, which remains liquid and therefore becomes richer in A.

In addition, the phenomenon of physical sorption is bivariant. The equilibrium equations of the sorption pair connect pressure, temperature and mass fraction, as indicated in Chapter 1 and as we will see in detail in section 2.3.

2.1.2. *Components of the one-stage AHP cycle*

The basic (or one-stage) heat pump system contains four components (Figure 2.1):

– a generator (also called a desorber) in which the poor solution is heated in order to allow for desorption of the working fluid (direct meaning of equation [2.1]);

– an absorber, receiving the rich solution from the desorber. This absorber is cooled in order to allow the exothermic absorption of the working fluid by the solution (direct meaning of equation [2.1]);

– a condenser in which vaporized working fluid from the desorber condenses, rejecting a quantity of heat which includes, mostly, the latent heat of condensation of the working fluid;

– an evaporator in which the working fluid coming out of the condenser evaporates due to the effects of a heat supply.

The four components (absorber, desorber, evaporator, and condenser) operate at two pressure levels and are generally connected by a pump and two expansion valves (Figure 2.1). The desorber and the condenser are at high pressure, while the other two components are at low pressure. The four phenomena in the four components take place simultaneously, and this system can therefore function continuously, as long as the external heat sources and heat sinks necessary for the different phenomena are available.

One of the main criteria used for the analysis of absorption processes is the thermal COP (coefficient of performance) (COP_{th}), which is defined in equation [2.2]. This criterion defines the efficiency of a process. In the case of absorption-based thermal machines, the relative share of electricity consumed compared to the heat used by the system is theoretically low, since the only electricity necessary for the operation of the system is that consumed by the pump located between the absorber and the desorber.

$$COP_{th} = \frac{\text{Useful heat (or cold) produced by the system}}{\text{Heat "consumed" by the system}} \qquad [2.2]$$

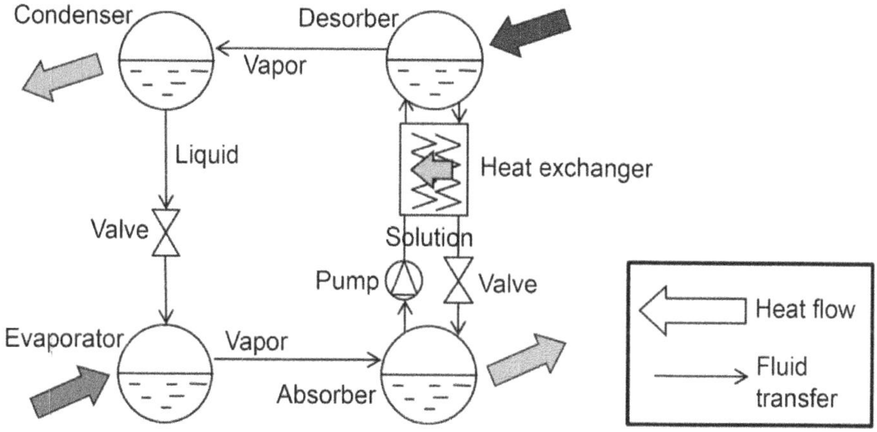

Figure 2.1. *Operation of the liquid/gas absorption-based heat pump system*

To improve the COP of the AHPs, many systems include an internal heat exchanger between the tubes of the rich and poor solutions, between the desorber and the absorber (Figure 2.1). This means that the solution exiting the desorber can transfer its heat to the cooler solution arriving from the absorber. Thus, the amount of heat required for desorption decreases and the COP of the system is improved (Srikhirin et al. 2001).

The typical COP_{th} of AHPs for the production of cold is on the order of 0.6–0.7 (Herold et al. 2016).

As explained above, the operation of these AHPs generally requires the use of a pump to circulate the solution (Figure 2.1). The other criterion for these machines, and the one which makes it possible to position the absorption machines relative to the vapor compression systems, will be referred to here as the electric COP (COP_{elec}), defined using equation [2.3], and sometimes also called the EER (energy efficiency ratio):

$$COP_{elec} = EER = \frac{\text{heat (or cold) produced by the system}}{\text{electricity consumed by the system}} \qquad [2.3]$$

Theoretically, the power consumed by AHPs is very low (on the order of 0.01% of the heat transmitted to the desorber by the working fluid "water"; Herold et al. 2016), which justifies the interest of these systems compared to vapor compression heat pumps, since the corresponding theoretical COP_{elec} values would be on the order of 100. There are also AHPs that do not use pumps (known as a "diffusion" system), but their thermal performance is more limited, and they are not used in the case of heat storage systems.

2.1.3. *Operating conditions of AHPs*

Representing absorption cycles often relies on pressure-temperature diagrams (generally represented in the form of ln(P)–(−1)/T), and referred to as Clapeyron, Oldham, or Dühring diagrams, as presented in Chapter 1. The single-stage heat pump cycle is shown in Figure 2.2, considering the system as exchanging heat with three different heat sinks or heat sources at three different temperatures T_b, T_0, and T_h. The equilibrium curves of the sorbent with a given mass fraction x (isoteres) are noted Sorb/G in Figure 2.2, using dotted lines. For the solution, we can therefore see that two of the operating parameters for equilibrium (pressure, temperature, or mass fraction) must be known in order for the third to be deduced. As mentioned in section 2.1.1, the behavior of the solution is therefore bivariant. The liquid/vapor equilibrium curve of the working fluid is represented as L/G in solid lines in Figure 2.2. In this case, since the working fluid is pure, only one of the equilibrium parameters needs to be known (pressure or temperature) in order to deduce the second one. The behavior of the phase change in the state of the working fluid is monovariant.

In the case of cooling, the most useful effect is the evaporation of the working fluid at a low pressure p_b and low temperature T_b (Figure 2.2). This phenomenon takes place in the evaporator (Figure 2.1). In the case of heat production, the two useful effects are condensation and absorption, rejecting heat at T_0 (above ambient temperature). These phenomena take place in both the condenser and in the absorber.

The liquid–vapor thermodynamic equilibrium positions of the working fluid (L/G) and of the sorption phenomena (Sorb/G) shown in Figure 2.2 thus create two distinct working pressures in the system: the high pressure p_h of the desorption phase, and the absorption at the lower pressure p_b. The two phases are simultaneous, with the sorbent solution being continuously recirculated between the absorber and the desorber (Figure 2.1). Thus, to allow the operation of the cycle, a heat supply at temperature T_h is necessary at the desorber.

It is also possible to operate this cycle for the production of heat at high temperature (T_h) from heat at intermediate temperature (T_0) by reversing the pressures of the phases, and imposing a desorption at low pressure and at T_0, and an absorption at high pressure and at T_h. This cycle (known as a "thermotransformer") is used on industrial sites for purposes such as recovering waste heat, though in a limited number of applications (Herold et al. 2016).

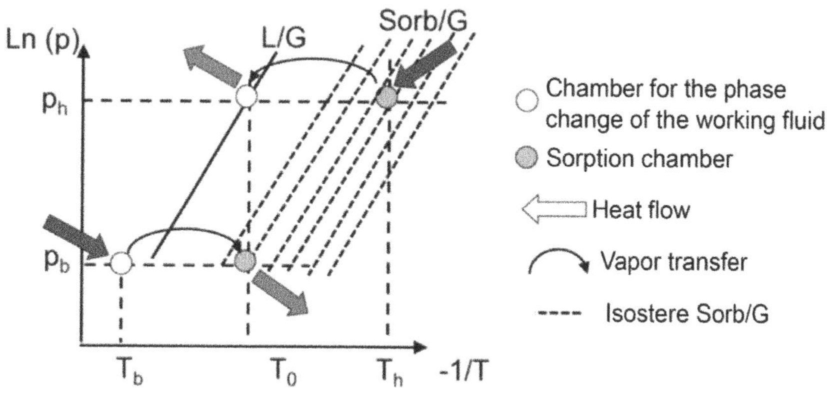

Figure 2.2. *Representation of the single-stage heat pump cycle (condensation and absorption exchanging with the same source at T_0) on an Oldham diagram*

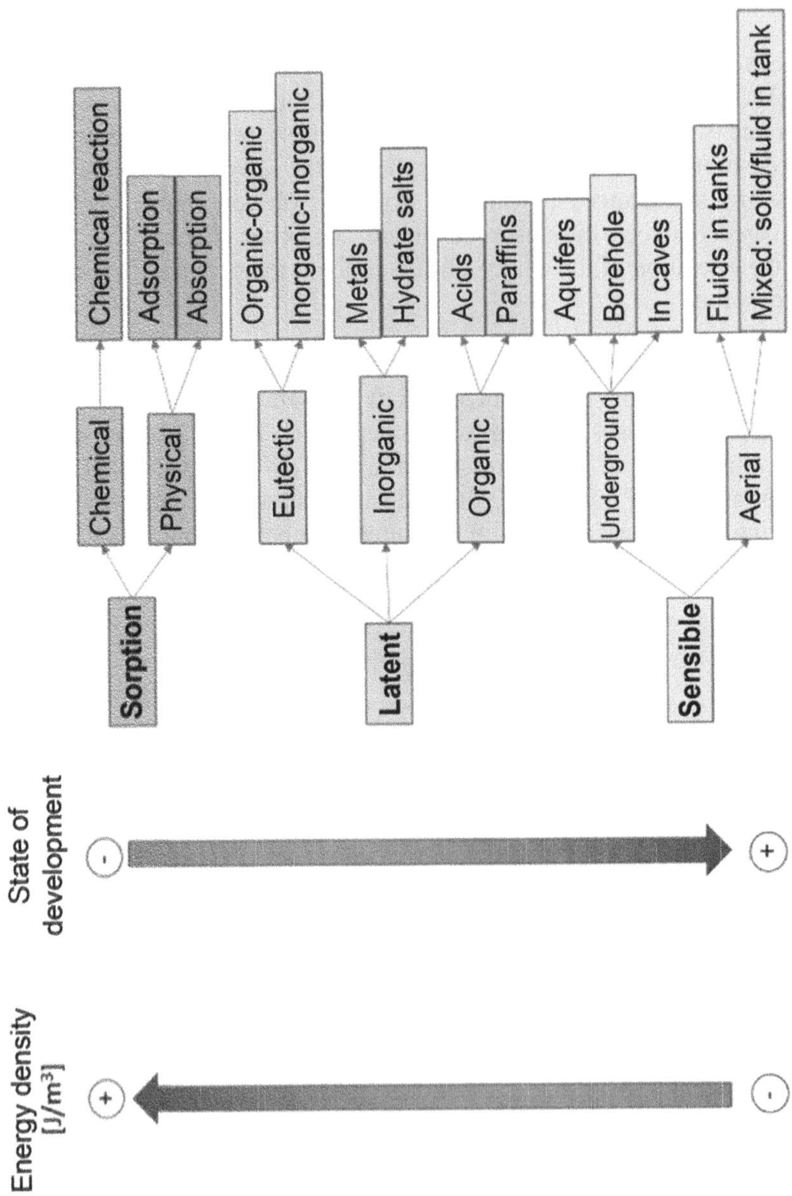

Figure 2.3. *Heat storage technologies, energy density and state of development*

2.2. Methods for storing heat by absorption

Absorption heat storage systems are based on the single-effect AHP cycle described in section 2.1. As with the AHP cycle, the storage of thermal energy by single-effect absorption has been developed for various applications, such as building cooling, space heating, domestic hot water (DHW) production and the thermotransformation of heat. The absorption process makes it possible to store both cold and heat, and the cold and the heat can be supplied simultaneously or independently. This makes the storage of thermal energy by absorption more flexible than conventional heat storage by sensible or latent heat.

This field of research is relatively new. The main projects concerning these processes have been developed over the last 20 years. The first developments concerned the inter-seasonal storage of heat. The concept of storage by absorption essentially demonstrates the advantages offered by a high energy density storage system as compared to sensible or latent heat systems (Figure 2.3). In addition, since the heat is stored in the form of a chemical potential in this case, the storage can therefore be carried out at ambient temperature. Consequently, the storage time has little influence on the COP of the system, which in this case is defined as the ratio of the amount of heat or cold discharged from the process to the amount of heat that has been previously loaded into it (equation [2.2]).

2.2.1. *Design of a system for storing heat by absorption*

As noted earlier, the principle of the method for absorption-based heat storage is based on AHPs. We utilize the fact that two components of the heat pump consume heat (the desorber and the evaporator), and two components reject it (the condenser and the absorber) to manage heat between periods of excess heat and times when the user needs heat. However, unlike the case of the AHP, the operation of the storage system is discontinuous in principle, since the heat charging periods (or the recharging or storage regeneration periods) are separated from the storage periods as such (periods during which neither removal nor addition of the energy occurs) and the periods of discharge (or use) of thermal energy. In addition, this storage method requires fluid storage tanks in addition to the components contained in the AHP.

Several configurations can be used for this system, requiring varying numbers of components, as shown in Figure 2.4.

The configuration closest to that of the AHP (Figure 2.4(a)) has three tanks to separately store the rich solution (i.e. the solution with a high mass fraction of sorbent), the poor solution (the solution with a low mass fraction) and the working fluid.

It is also possible to envision an intermediate system (Figure 2.4(b)), which requires two fewer exchangers. In the storage process, the absorber and the desorber do not work simultaneously, since the system is either in the charging phase (implementing the desorber and condenser) or in the discharging phase (implementing the absorber and evaporator) or in the storage phase (in transition between charging and discharging, where no exchanger is implemented). Thus, the desorber can be designed to function as an absorber as well. In such a case, it is referred to as a "reactor" (Figure 2.4(b)). Likewise, the condenser can serve as an evaporator.

A third, more compact configuration (Figure 2.4(c)), and therefore one that is a priori less expensive, is similar to the previous configuration, but with a single solution storage tank. The solution is then recirculated there, and the mass fraction in this reservoir changes gradually over the course of the charging and discharging. This configuration was proposed and developed in (N'Tsoukpoe 2012). It is this configuration that will be described in the following sections of this chapter.

Another absorption storage configuration that is even more compact has been proposed by Quinnell and Davidson (2014) and is presented in Figure 2.4(d). The basic operating principle is the same as that of Figure 2.4(c), but the solution and the working fluid are stored together in a single tank. It is possible as the solution has a greater density than the working fluid, which becomes even greater as the richness of the solution increases. We therefore have a pseudo-stable stratification of mass fraction inside the tank.

a)

b)

c)

d)

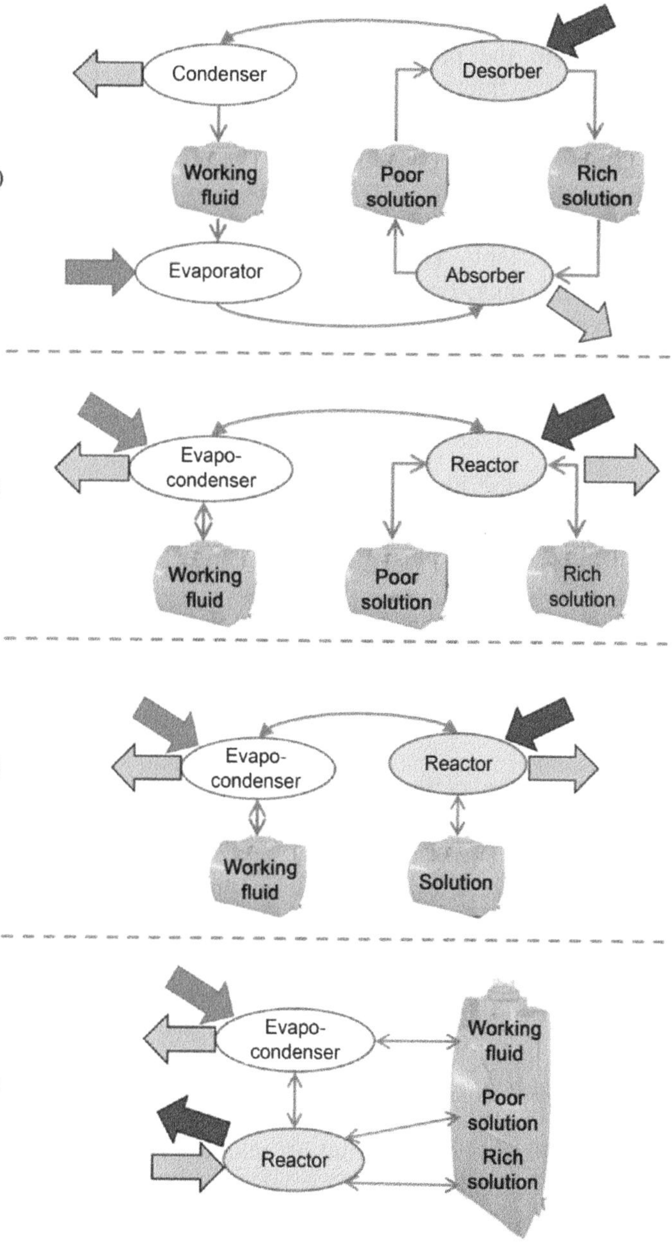

Figure 2.4. *Possible configurations for the absorption heat storage process: (a) base configuration; (b) intermediate configuration; (c) compact configuration; (d) integrated configuration*

Nevertheless, the natural phenomenon of diffusion, even if it occurs very slowly compared to the phenomenon of convection, tends to gradually homogenize the mass fraction of the sorbent throughout the reservoir, and for storage periods greater than a few weeks, a decrease in this stratification is observed. In addition, the impact of the temperature on the density (with density of the fluid becoming greater as the fluid becomes colder) greatly complicates the phases of addition and extraction of the solution and working fluid in the tank. The two fluids must be injected at exactly the correct levels in the tank, depending on both their mass fraction and their temperature, to avoid a natural convection which would destratify the layers of the mass fraction. This configuration, that is in principle the most compact but also the most difficult to implement, will therefore not be further developed over the rest of this chapter.

When looking for the best compromise between the energy density, the storage period and the simplicity of the design and operation, it is the compact configuration presented in Figure 2.4(c) which is used. The two exchangers (sorption reactor and working fluid phase change exchanger) can be positioned in the same tank, so as to further densify the process and avoid pressure losses related to the vaporized working fluid flow between these two exchangers. The design of the absorber/desorber sorption reactor, a key component of the system, will be detailed in section 2.3.

The operation of the system during the different phases (charging and discharging) for the compact configuration is presented in Figure 2.5, using the temperatures of an interseasonal solar heat storage unit for the heating of the building as example temperatures. The method operates in "charging" mode in the summer, when solar heat is abundantly available, and in "discharging" mode in winter, when heating is needed. Here, the heat production temperature represented corresponds to a use in a low-temperature heating floor.

During charging, heat flux at temperatures of about 70–100°C, which are easily reached by solar thermal collectors, is delivered to the poor solution circulating in the desorber. The solution releases the working fluid in the form of steam and becomes richer. This working fluid is condensed inside a condenser through the release of heat into the atmosphere, which we assume to be at 30°C under summer conditions, and then stored in a tank. The rich solution is recirculated to its storage tank. This operating phase corresponds to the high-pressure part of the heat pump cycle (Figure 2.2).

The two storage tanks can simply be separated during the inert storage phase by closing a valve between the reactor and the evapo-condenser, stopping the mass exchange, and the energy can then be stored in the form of a chemical potential for as long as desired.

In the winter period, when heating is needed (discharging), the valve is reopened and the rich solution can absorb the vapor of the working fluid into the absorber, which is evaporated from a low temperature heat source (for instance, here it is assumed to be about 5°C) (Figure 2.5). The heat released by the exothermic phenomenon of absorption is used for heating the building. This operating phase corresponds to the low pressure part of the AHP cycle (Figure 2.2).

In this example, it can be seen that the temperatures of the reactor during the two operating phases are not equal, and therefore that the temperature of the heat consumed by the charging is different from that which can be recovered in the discharging phase. These temperatures depend on several parameters, as will be described in section 2.2.2. They must be adapted to the heat source and the needs considered.

Storing heat in the way described above is one possible use of this absorption storage method, but it also allows for storing colder temperatures for the cooling of spaces. For this, the operation of the cycle is very similar to that already presented, but it differs in the connection of the exchangers with the user in the discharging phase. In this case, it is the heat necessary for the evaporation of the working fluid in the evaporator that will be consumed by the system, which then allows for useful cooling. This allows the system to produce colder temperatures at about 5°C, as shown in the example in Figure 2.5. The heat removed by the absorber is transferred to the outside environment during this period. This use makes it possible to shift a period of high heat availability from a period when cooling is needed, for applications such as certain industrial uses. Nevertheless, in buildings, the periods when cooling is needed and those with high levels of solar flux are generally simultaneous; in this case, a simple AHP without storage is more suitable.

One differentiating characteristic of this storage method with respect to sensible or latent storages or other thermochemical storages is that it intrinsically dissociates the storage and charging/discharging functions: the storage takes place in the tanks, while the charging/discharging occurs in the reactor. Thus, each component can be dimensioned separately based on:

– the amount of heat to be stored during the cycle for the volume of the tanks;

– the power to be used while charging and/or discharging for the reactor (circulation flow rates, exchange surfaces).

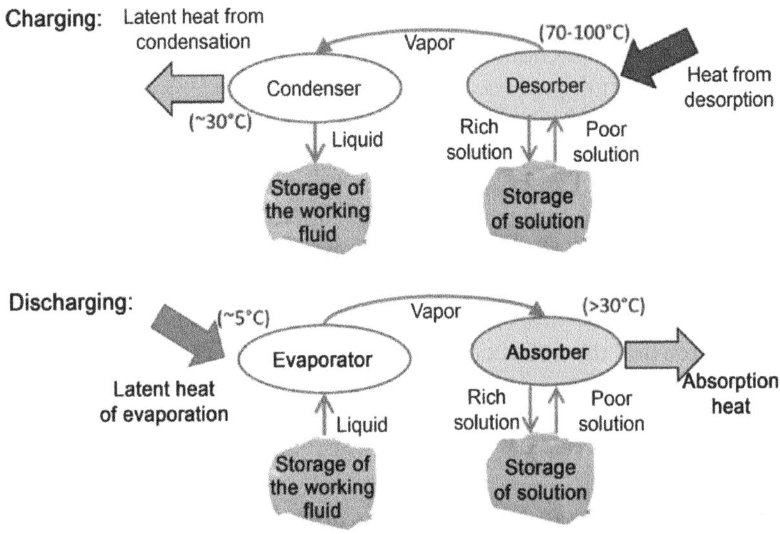

Figure 2.5. *Principles of the operation of the heat by absorption storage method*

2.2.2. *Operating conditions*

An example of an annual thermodynamic cycle associated with the compact configuration, in the case of the LiBr–H_2O work couple (see Chapter 1 for more information regarding the choice of the sorption couple), is shown in the Dühring diagram shown in Figure 2.6.

At the beginning of the charging period, the poor solution (e.g. 48 m% of LiBr) is at the temperature of the storage tanks, which in this case are assumed to be at the soil temperature of 5°C (point 1a). When heat inputs are available, this poor solution is pumped from the solution tank to the desorber. It is then heated and increases in pressure until the liquid/vapor equilibrium is obtained in the condenser (point 2a). If the heat supply continues and the temperature level is sufficient, the solution releases water vapor and is enriched (reaching the state noted 2b at 57 m% LiBr, for instance), while the vapor condenses in the condenser (point 3a, at 30°C in this case): this is the effective phase of desorption. The liquid water leaves the condenser and accumulates in the water tank, where it gradually cools down to the storage temperature (point 4, at 5°C). Meanwhile, the LiBr-rich solution that comes out of the desorber circulates toward the solution reservoir, where it gradually cools down to the storage temperature (point 1b, also at 5°C.).

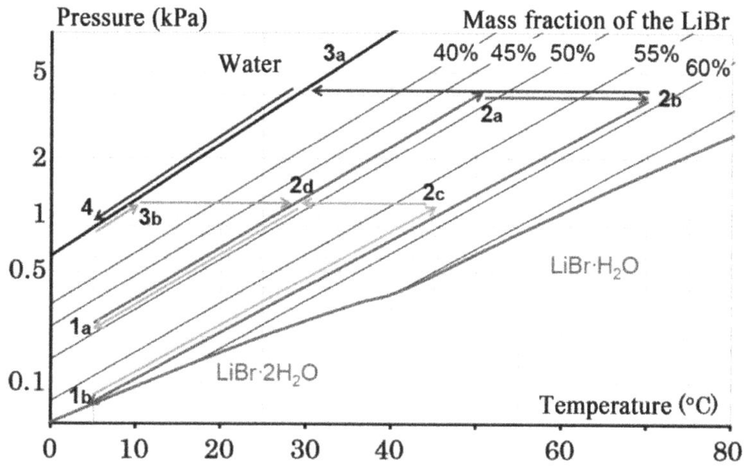

Figure 2.6. *Annual thermodynamic cycle of the heat storage process in the Dühring diagram for LiBr-H_2O (N'Tsoukpoe 2012)*

As the charging progresses, the mass of the solution in the solution tank decreases, while its mass fraction increases. The same occurs for the desorbed and condensed water that accumulates in the water tank. By blocking the mass transfer between the solution tank and the other components of the process, this sorption heat can be stored with low, in principle, thermal losses, since the energy is stored in the form of a chemical potential. Only the sensible heat is lost during the cooling phase 2b → 1b, if it is not recovered by a solution exchanger as shown in Figure 2.1.

During times when heating is needed (discharging phase), the rich solution is transferred to the absorber, while the water is pumped from its tank to the evaporator where it evaporates (point 3b, Figure 2.6) at a low temperature. The heat necessary for this evaporation at a low temperature is provided by the environment, at temperatures such as 10°C for example. The water vapor that is produced in the evaporator is absorbed by the solution until reaching the pressure equilibrium in the generator and the evaporator (point 2c at 45°C). The absorption takes place at a constant pressure (2c → 2d), and the heat produced by this exothermic phenomenon can be used for heating. The poor solution that is obtained (point 2d) returns to the solution reservoir (point 1a) where it cools, after having previously transferred its heat to the solution entering the absorber, using a solution exchanger as shown in Figure 2.1.

In cases where cooling is needed, the heat rejected by the absorber during the discharging phase is transferred to the environment. In this case, the temperature of the absorber may be lower than in the case of space heating applications. Consequently, the pressure of the discharging phase can also be lower and the evaporation temperature can also be slightly lower than in the case presented above, making it possible to provide useful cooling to the user.

In this example of the couple of LiBr–H_2O, as in all the processes of heat storage by absorption developed up to this point, the working fluid is water. The choice of this fluid is based on the criteria of very high availability, minimal cost, a complete lack of risks and being completely safe for humans and their surrounding environment, as well as its energy performance (in particular, a very high latent heat), as described in Chapter 1. Nevertheless, a major drawback of this fluid is the pressure that it requires in the system under its operating conditions for low temperature storage: for evaporation/condensation temperatures generally between 5 and 40°C, the pressure in the system is 0.8–9 kPa (absolute). The system must therefore operate in a partial vacuum relative to atmospheric pressure; a significant constraint on its technological design, as will be discussed in section 2.3. The other limit in the case of the working fluid "water" is, of course, a temperature requirement of at least 0°C at the evaporator to prevent the water from freezing.

The system has a certain degree of flexibility in its operation. During both the charging and discharging periods, the process naturally adapts both its pressure and temperature to the external sources to which it is subjected. For example, in the charging phase, if the heat source temperature is at a higher temperature, one of these two things can occur:

– either the pressure in the system will be greater and the condensation will take place at a higher temperature;

– or the condensation is maintained at the same temperature and therefore the same pressure in the identical system, and the mass fraction of the solution can be increased to a higher level, for example, 61% for a heat source at 75°C and the sorbent LiBr (Figure 2.6).

Similarly, if the available source temperature for evaporation is only 5°C during the discharging phase, evaporation-absorption is possible and is carried out at a lower pressure, but the absorption temperatures will only allow a maximum of 40°C to be reached in the absorber.

2.2.3. *Performance indicators*

Some energy performance criteria for this storage method are identical to those used for AHPs: the thermal COP, as described in equation [2.1], both for heat and cold storage, and electrical COP as described in equation [2.2]. Here, as with heat pumps, the thermal performance of the system is highly dependent on the charging and discharging conditions, but also on the duration of the storage period. The longer this period lasts, the more the solution tanks and working fluid will effectively lose heat to the outside, and therefore the more the sensible heat that accumulates in the tanks during the charging period is lost, which tends toward the case shown in Figure 2.6, with the two tanks at the soil temperature at the beginning of the two phases. In this case, only the chemical sorption potential remains stored in the process. In the case of a shorter storage period, the temperature of the fluids does not drop as steeply and can be utilized during the next useful phase. Depending on the storage period, the theoretical thermal COP of the system therefore varies between approximately 60% and 85% (N'Tsoukpoe 2012). Thus, in the most unfavorable case, where the storage period is long enough so that all of the sensible heat is lost in an inter-seasonal type of storage, during the discharging it is possible to recover a minimum of 60% of the heat put into the system during charging. This value increases for shorter storage times.

Another essential performance indicator in heat storage is energy density. This indicator quantifies the specific stored heat, which is to say the heat per unit mass of the system or per unit volume of the system. The choice of which quantity to use depends on which constraint will be the most significant for the projected application. Typically, for static systems used in applications such as heating for buildings or industrial use, the most important constraint is generally the size of the storage, and therefore volumetric storage density is used. Here, one specific consideration must be taken into account to quantify the volume of this process compared to sensible or latent heat storage technologies: for other technologies, it is often sufficient to take into account the volume of storage material used, since it sees very little variation during operation and it does not circulate. In the case of absorption storage, by using this definition, we will be taking into account the volume of the poor solution at the end of the absorption phase. At this point in the cycle the water tank is empty, and therefore the volume of the solution is at a maximum. However, this definition of the volume of the system is very favorable, as the system involves both a solution tank and a working fluid tank and one (or two) containers containing the sorption reactor and the "evapo/condenser". These elements should also be taken into account in determining the volume of the system and therefore in calculating the energy density of the process. Thus, the volume density of the storage varies greatly for this process, which will depend on the amount of heat to be stored by the system: when the system is used to store a few kilowatt-hours, the volume of the container is the same order of magnitude as that of the tanks and is detrimental to the observed energy density. Therefore, for much larger amounts of storage, the actual energy density will be greater. Different definitions of storage densities can be found in the scientific literature, whose use depends on the case in question. The energy density of the absorption storage, which also varies depending on the sorption couple that is chosen, is on the order of 150 kWh/m^3 on the scale of the overall process (Stutz et al. 2017), which is about two times lower than if we refer to the volume of solution alone, but remains much larger than a sensible or latent heat storage process.

One way to take into account the different forms of energy involved in a system is by using the criterion of exergetic efficiency or effectiveness η (equation [2.4]). This criterion takes into account both the quantity and the quality of the energy involved (Descombes and Desmet 2022).

$$\eta = \frac{\text{Exergy of the useful heat (or cold) produced by the system}}{\text{Exergy consumed by the system}} \qquad [2.4]$$

In an ideal solar heat storage system which makes it possible to produce as much heat or cold in the discharging phase as that consumed in the charging phase, the theoretical exergetic efficiency of the heat storage system is 1. However, significant irreversibilities occur in the process, particularly due to the great difference between operating conditions in summer versus winter. The exergy linked to the heat exchanged by the system can be quantified from a cycle analysis, as shown in Figure 2.7.

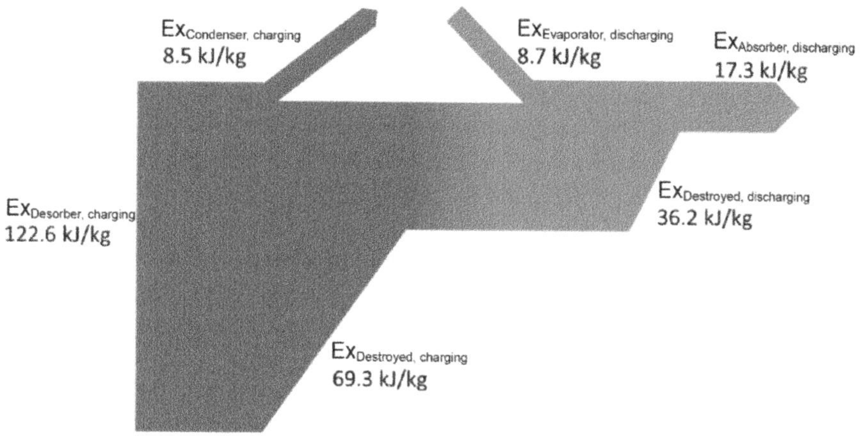

Figure 2.7. *Exergy exchanged by the system for the absorption storage cycle with the couple of LiBr–H_2O (the exergy is quantified in reference to the mass of LiBr) (Perier-Muzet and Le Pierrès 2016)*

The detail of this method is presented in N'Tsoukpoe et al. (2014) and in Perier-Muzet and Le Pierrès (2016), examining the case of an inter-seasonal storage system with summer and winter operating conditions similar to those presented in Figures 2.5 and 2.6. The COP of the cycle calculated under these conditions is 0.49. Also under these conditions, the volumetric energy density for the storage system is 73 kWh/m^3, quantified here with reference to the volume of the two storage tanks (containing the solution and water).

The exergy consumed by the cycle at the level of the desorber is 122.6 kJ/kg$_{LiBr}$ and expressed with reference to the pure sorbent put into use (Figure 2.7). The condenser exchanges 8.5 kJ/kg$_{LiBr}$ of exergy-heat with the outside environment during the charging phase. During the cycle, heat is also consumed in the evaporator during the discharging phase, equal to 8.7 kJ/kg$_{LiBr}$ of exergetic content. The heat produced by the reactor (absorber) during discharging (here, the useful effect) represents an exergy of 17.3 kJ/kg$_{LiBr}$.

The thermal losses related to the sensible heat of the solution and the water in the reservoirs lead to the destruction of exergy. Similarly, the heat transfers in the exchangers, which are not supposed to be ideal here, also lead to exergetic destruction. The exergy lost through the various irreversible aspects of the system represents a total of 69.3 kJ/kg_{LiBr} in the charging phase and 36.2 kJ/kg_{LiBr} in the discharging phase, during which the pinches in the exchangers are lower and therefore the irreversible phenomena linked to the thermal transfers are lower. However, in the case of an interseasonal storage system, it is impossible to reduce a large part of this unused energy, which is more closely related to the significant shifts in the ambient conditions from summer to winter, depending on the operating phases of the system rather than on the intrinsic operation of the process.

In this instance, the exergetic effectiveness of the cycle can be calculated at 0.15, if we take into account the irreversibilities of the transfers during exchanges with the sources, or at 0.22 if we only look at the internal irreversibilities. The exergetic efficiency of the cycle is lower than the thermal COP, since the useful heat produced by the system during the winter (at a temperature around 35°C) is of a lower quality than that consumed by the system in the summer (at a temperature above 60°C). However, the process (if it is operated, for example, over short storage periods with an almost constant ambient temperature between the two phases) would have a good intrinsic exergy efficiency.

2.3. Reactors

Among the components of absorption processes, the exchangers of the working fluid (the evaporator and condenser) have been extensively studied, since technologies for evaporators and condensers are widely used in multiple industrial applications and physical phenomena.

In the case of absorption storage where the working fluid used is the operating pressure of the cycle is lower than atmospheric pressure, and therefore very close attention must be paid to the hermeticity (tightness of the seal) of the system. Leaks can lead to the entry of air into the system, which accumulates and remains in the form of vapor (which is then known as "incondensable" in the operating conditions of the cycle), which significantly reduces the performance of the exchanges, particularly in the condenser. In addition to this risk of leakage, these two components do not have any specific particularities.

If no incondensables are present, transfers in the vapor phase are not the limiting phenomena (Herold et al. 2016). On the other hand, numerous studies have shown that the phenomenon of absorption/desorption is controlled by the transfer resistance (in particular the mass transfer) in the liquid phase. As noted previously, in the case of systems operating with water as a working fluid, the pressure in the system is very low. In this case, the density of the vapor phase is also very low. This makes it necessary to maintain sections for relatively large vapor phase flows to pass (since this vapor must circulate efficiently between desorber and the condenser, as well as between the evaporator and the absorber). Consequently, the exchangers developed for the absorber and desorber in this case must have a liquid/vapor interface that is as large as possible to allow for the sorption phenomenon to take place at this interface, and for vapor to flow easily. The flows in these exchangers thus occur in a film (Figure 2.8).

In the case of film absorbers and desorbers, a solution dispenser is located at the top of the exchanger to allow the solution to flow downwards on the heat exchange surfaces with the external heat transfer fluid (walls). The film formed this way is where both the heat transfers and the mass transfers occur. To illustrate these phenomena, we will examine the case of absorption. In an absorber, the working fluid is absorbed at the liquid–vapor interface, which leads to a decrease in the mass fraction of the sorbent. Then, due to the phenomenon of diffusion and/or convection in the film, this working fluid mixes over the entire thickness of the solution film. In the case where this mixing is ineffective (which may be the case for laminar flows), the mass fraction remains low at the interface, and since the pressure in the system is established by the pressure in the condenser, the equilibrium temperature of the solution at the liquid/vapor interface decreases. Consequently, in order to allow the absorption phenomenon to be maintained, this interface must be cooled down to a lower temperature. Since the temperature of the external heat transfer fluid which receives the heat transfer from the absorber cannot continuously decrease, this tends to slow down the absorption of additional steam. It is therefore appropriate to intensify the mass transfers in the film.

At the same time, the heat released by the absorption phenomenon at the liquid–vapor interface must be transferred through the liquid phase to the wall of the exchanger. An increase or decrease in temperature at the interface will effectively lead to a change in the operating conditions, and therefore also tend to slow down the sorption rate. It is therefore appropriate to intensify the heat transfers in the film, which would encourage the thickness of the film to be reduced in order to reduce its thermal resistance. This shows that the heat and mass transfers that happen in the film are complex and interrelated phenomena. They are influenced by the flow regime of the solution, which can take different forms such as drops, columns, or entire layers of films, depending on the flow rate of the solution and the shape of the wall of the exchanger. The shape of the solution flow is also directly correlated with the wettability of the exchange surface.

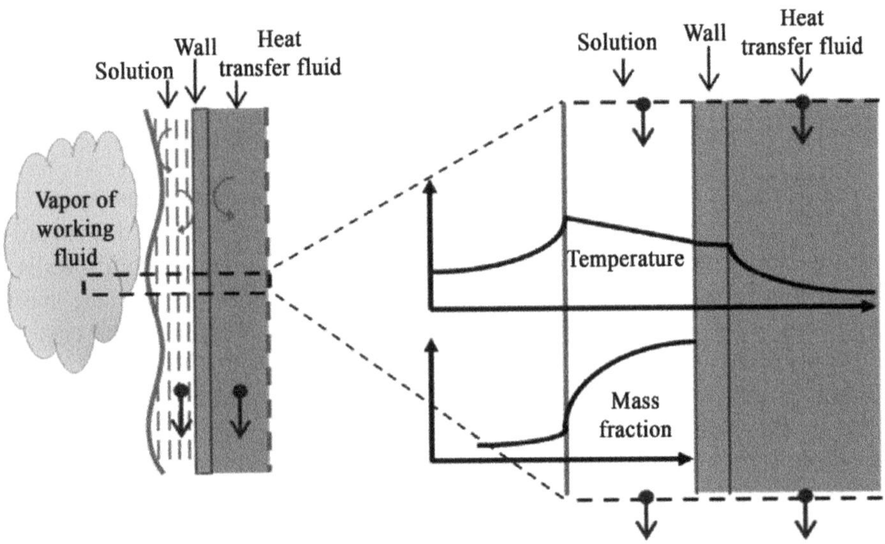

Figure 2.8. *Principles of operation and transfers in an absorber*

In addition, as has been detailed previously, one of the goals of this system is for the sorption reactors to be reversible, or in other words, for them to function both as absorbers and desorbers depending on the phases of the cycle. Moreover, the heat transfers on the side of the external heat transfer fluid can also be limiting (Herold et al. 1996), which means that a good overall design of the reactor is essential for the proper functioning of the system. Various reactor designs have been proposed in the scientific literature over the past decades, which are classified here according to their tube or plate geometry.

2.3.1. *Absorption reactors using tubes*

The shell and tube exchanger technology (Minvielle et al. 2014) is widely used and is most commonly found in AHPs using water as a working fluid (Altamirano et al. 2020). This has made it a natural fit for use in absorption storage. The exchange surface can be a horizontal or vertical tube. Horizontal configurations have historically been the predominant technology in commercial AHPs because of their versatility and simple operation. In the case of bundles of horizontal tubes, the instabilities that allow for the proper mixing of the film solution are generated during the drainage of the solution from one tube to another (Figures 2.9(a) and (b)), while for vertical tubes, the instabilities are generated by the waves which form on the surface of the film during its flow (Figures 2.9(c) and (d)).

For horizontal tubes (Figures 2.9(a) and (b)), the film flows outside the tubes and the heat transfer fluid flows inside them. The formation of the drops and their impact on the tubes generate instabilities; the frequencies of these depend on the flow rate of the solution (the Reynolds number), its thermophysical properties (density, viscosity and surface tension) and the distance between the tubes. These instabilities are beneficial to the transfer phenomena, because they make it possible to increase the surface of the interface, to strengthen the convective heat and mass transfers in the film, and to refresh the exchange surface of the solution (Altamirano et al. 2020).

For vertical tubes, the falling film can be located outside (Figure 2.9(c)) or inside the tube (Figure 2.9(d)). Various studies conducted on this configuration have shown that when the tubes are textured (using fluted, grooved, or twisted textures, for example), the absorption performance of vertical tubes is close to that obtained in horizontal tube absorbers. An optimal tube length can be determined, beyond which the absorption rate of the exchanger remains unchanged. Vertical tubes with an internal flow of the solution are of particular interest in the case of absorbers cooled using air as the heat transfer fluid due to the fact that fins can easily be added to their outer surface.

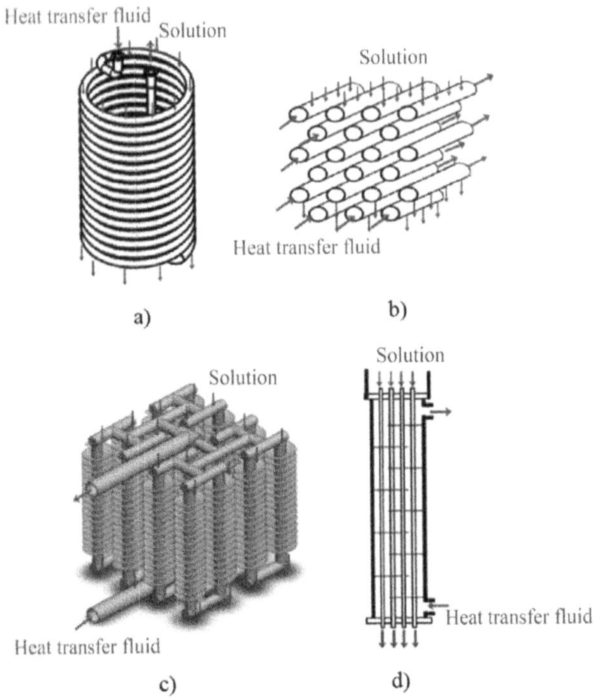

Figure 2.9. *Configurations of exchangers used in absorption storage processes: (a) using helical horizontal tubes; (b) using parallel horizontal tubes; (c) using vertical tubes with external film; (d) using vertical tubes with internal film*

In the case of a storage, in order to reduce both the electrical consumption used to circulate the solution as well as the transfer resistances in the films as much as possible, a very low solution flow rate in the exchangers is sought. The film flow regimes are therefore laminar (with a Reynolds number less than 10) or transition (with a Reynolds number between 10 and 500). Several prototypes and devices that have been developed have implemented this technology. We can find both horizontal tube exchangers with a helical arrangement, as developed by Bales and Norlander (2005), or more recently with parallel horizontal tubes, as in the device created by Zhao et al. (2016) using the couple LiCl–H_2O (Figure 2.9a), as proposed by Zhang et al. (2014) with the couple LiBr–H_2O, Daguenet-Fick et al. (2016) with the couple NaOH-H_2O or Gao et al. (2020) with the couple of LiBr-H_2O (Figure 2.9(b)), or vertical tubes developed by Liu (2011) with the flow of the external solution on helical fins for the pairing of $CaCl_2$-H_2O

(Figure 2.9(c)) or N'Tsoukpoe et al. (2013) with the internal flow of the solution for the couple LiBr–H_2O (Figure 2.9(d)). The difficulties these authors faced are different in nature: in particular, problems were encountered in wetting the wall of the tube, which reduced the surface area of the liquid–vapor interface and therefore the performance of the exchanger, and the complexity of the distribution of the fluid over all of the tubes.

2.3.2. *Absorption reactors using plates*

Another exchanger technology has also emerged: plate exchangers. Commercial heat exchangers using plates can easily operate using falling film at low flow rates. The main advantages of these are their low cost, flexibility, compactness, ease of maintenance and good performance. They can be joined with joints (which makes them easier to clean and maintain but is not recommended in the case of operation under a vacuum due to the possibility of leaks) or brazed. The exchange surface can be smooth or corrugated. In this second case, the flow instabilities generated improve the transfers, as has also been noted in the case of tube exchangers. As with tube exchangers, particular attention must be paid to the distribution of the solution on the plates for them to be able to be wetted properly, otherwise the performance of the exchanger is greatly reduced (Altamirano et al. 2020). One problem faced by plate exchangers is the fact that in the case of smooth plates, the mixing of the fluid is greatly limited with respect to the case of the driping between horizontal tubes.

Various options have been studied to improve the heat and mass transfers in falling films within plates. A configuration of horizontal plates (Figure 2.10(a)) allowing for the circulation time of the solution in the exchanger was tested by Weber and Dorer (2008) with the couple NaOH–H_2O. Then, grooved plates that allowed the film to be better distributed over the surface of the plate were tested by Huaylla (2017) and Le Pierrès et al. (2017) with the couples of LiBr–H_2O and KCOOH-H_2O, respectively (Figure 2.10(b)). The grooved plates are an interesting option since the grooves make it possible to obtain a stable wetting rate, even at low flow rates. However, the maximum exchange surface area is reduced, and the observed turbulence effects are lower compared to flows on smooth plates (Stutz et al. 2016). Overall, the results obtained with these plate exchangers in sorption reactors present the same challenges as tube exchangers, with the high difficulty of wetting the plates with a low Reynolds number, which can reduce the performance of the exchanger.

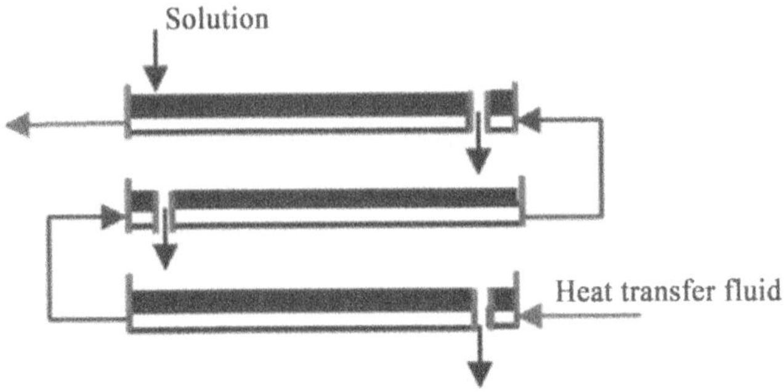

a)

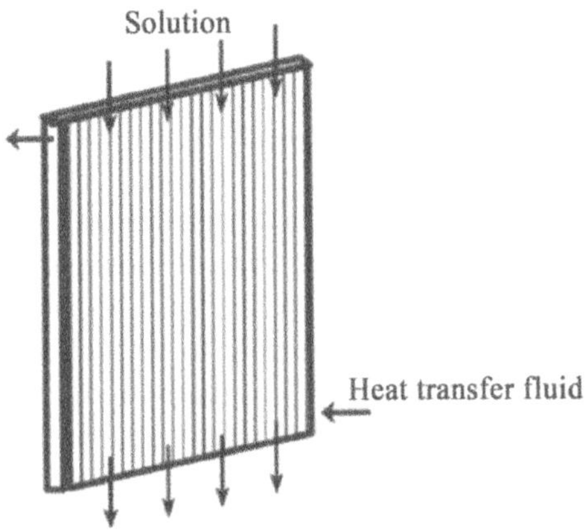

b)

Figure 2.10. *Configurations of exchangers seen in the processes of storage by absorption: (a) using horizontal plates; (b) using grooved vertical plates*

2.4. Intensified storage cycles

As explained above, some of the fundamental criteria for storage systems are the storage COP and the energy density of the system. Sorption processes have an inherent ability to store heat for theoretically infinite periods and a high energy density compared to sensible or latent storage processes. But these advantages come at the cost of an increased complexity of the system. As a result, various solutions have been proposed to further increase the energy density of the system and make the process competitive.

2.4.1. *Intensification through the crystallization of the solution*

During the discharging phase of the process (absorption phase), the larger the amount of working fluid absorbed by the solution, the greater the sorption heat. Therefore, during the desorption phase, which represents the charging of the process, the more working fluid is separated from this solution, the more energy is stored (Figure 2.6). In the case of an aqueous solution resulting from the mixture of water and a salt, if the mass fraction of the salt solution reaches the solubility limit for its temperature, the solution crystallizes partially or completely (the same behavior that is observed in salt marshes with the crystallization of sea salt when the amount of water decreases through the phenomenon of evaporation). Thus, the crystallization of the solution is a phenomenon that allows the energy density of the process to be increased. However, in general this phenomenon is specifically avoided in absorption processes, because of the risks that it entails with respect to the operation of the machine (risks of clogging of the pipes with salt crystals, damage to the pump, etc.). However, in the case of the heat storage process, it was shown that it is feasible and a promising new option. For this process, crystallization can be allowed at the level of the solution reservoir, since this component does not see a continuous circulation of the fluid. The related annual thermodynamic cycle for the pairing $LiBr–H_2O$ couple with crystallization is shown in Figure 2.11.

In Figure 2.11, the saturation curve of the salt solution is represented by the solid green line to the right of the figure. To the right of this curve, the condition for the equilibrium of the solution is the salt being in solid form, either in the form of LiBr monohydrate $LiBr\text{-}H_2O$ for temperatures above about 40°C, or in the form of LiBr dihydrate $LiBr–2H_2O$ for lower temperatures. For the case of the cycle presented here, the maximum mass fraction of the solution chosen was 62% (this point is indicated as 2b). The cycle is identical to the cycle without crystallization until the cooling phase of the solution within the reactor (step denoted 2b → 1b). During this phase, when the temperature in the reservoir reaches approximately 35°C, the solution is saturated and crystals appear. The amount of crystals then gradually increases as the solution continues to cool, up to point 1b. Part of the contents of the tank are therefore in the form of solid $LiBr\text{-}H_2O$ crystal and the solution, which still remains in liquid form, has a mass fraction of 58% (which has solubility at 5°C).

During the next discharging phase, the mass fraction of the solution pumped into the tank will remain constant at about 58%, as long as the solid is present in the tank. The solid crystals dissolve gradually by recirculation of the poor solution leaving the absorber as shown in Figure 2.5. When all of the crystal has been dissolved, the cycle will again be similar to the cycle without crystallization and the solution will be progressively diluted from point 1b to point 1a in the tank at 5°C.

The maximum acceptable mass fraction during the storage cycle depends on the nature of the crystal formed in the temperature range in question and on the minimum quantity of liquid that is desired to be present in the solution tank. In the example of LiBr, as mentioned above, the crystal form of the crystallized LiBr is either mono or dihydrated (Figure 2.11), for temperatures between 5 and 90°C. Consequently, part of the water in the tank is trapped in solid form in the crystals that are formed. The amount of solid phase in the tank must then take this crystallized water into account. For example, for a maximum mass fraction during the cycle of 62%, the mass content of solids in the tank at point 1b is 33%.

Modeling has made it possible to quantify the energy densification allowed by the crystallization of the solution.

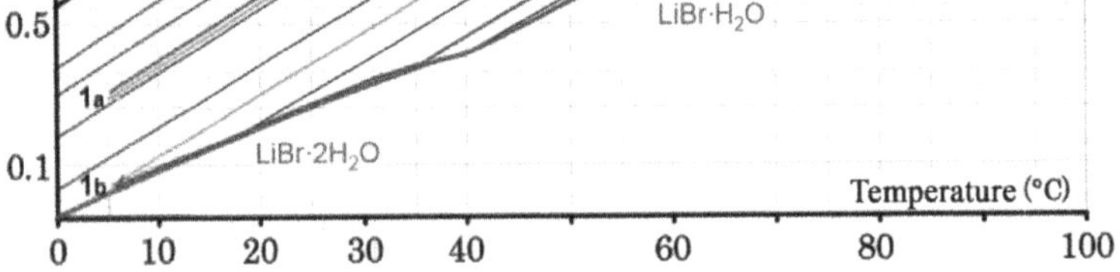

Figure 2.11. *Annual thermodynamic cycle of the heat storage process with the couple of LiBr–H$_2$O for a crystallization rate of 33% (N'Tsoukpoe 2012)*

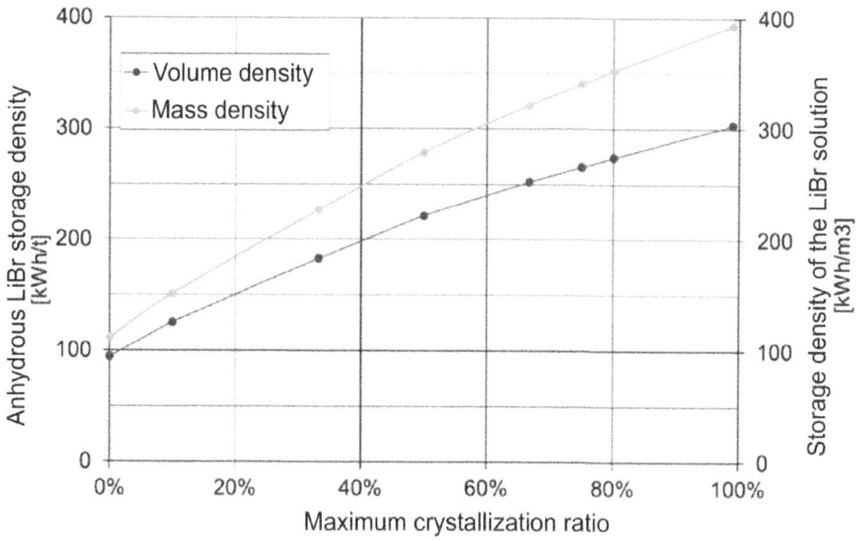

Figure 2.12. *Impact of the maximum crystallization ratio on the mass and volume storage densities for the LiBr–H2O couple (N'Tsoukpoe 2012)*

Figure 2.12 shows the energy density of the storage as a function of the maximum mass ratio of crystals in the solution tank. These mass and volume densities are calculated for the case of storage with LiBr-H_2O to cover the heating needs of a single family house. Crystallization can increase the storage density by more than three times, from about 100 kWh/m^3 to more than 300 kWh/m^3 under these operating conditions. However, depending on the technological constraints of operation, and in order to allow the circulation of the solution from the reservoir to the absorber in the discharging phase and then its recirculation to the reservoir for a progressive dilution of the solid, too high a level of solid in the reservoir should be avoided. It is estimated that above a maximum crystallization ratio of 50%, the technical complexity of the process increases (in terms of the circulation of the solution and managing the crystallization). Therefore, a maximum crystallization rate must be determined, with the goal of obtaining a workable compromise to achieve an efficient process. Similar results have been obtained for other sorption pairs, as presented in N'Tsoukpoe (2012).

2.4.2. *Intensification through modifying the cycle: two-stage cycles*

For AHPs, different cycles have been developed and implemented to improve the performance of one-stage systems. These improvements may seek to achieve a greater rise in temperature, a larger COP of the cycle or a greater flexibility of the system. For the same purposes, new heat storage cycles have also been studied in recent times. Additional components (heat and mass exchangers) are required in this case versus the one-stage cycle shown earlier. Two primary changes to the cycle have been proposed: using an internal heat transfer or an internal solution exchange.

2.4.2.1. *A two-stage cycle with an internal heat exchange*

It is possible for the one-stage cycle to be changed in order to mitigate the inconvenience of reducing the discharging temperature of the heat in relation to that used for charging (Gao et al. 2020). This decrease is related to the fact that the outside temperature available to the evaporator during the discharging phase is lower than that of the condenser during the charging phase. As a result, this situation is typically observed for interseasonal summer/winter heat storage. The two-stage cycle to alleviate this is presented in Figure 2.13.

The development of this two-stage cycle requires the combined use of two similar devices operating in parallel; the system is altered so that it consists of two absorbers/desorbers (referred to as 1 and 2 in Figures 2.13 and 2.14) and two evapo-condensers (also referred to as 1 and 2).

During the charging phase (Figure 2.13(a)), the two stages are separated and operate in parallel: as with the one-stage cycle, the poor solution leaving the tank where it is stored and circulated into the desorbers, where it desorbs the working fluid which is condensed in the two condensers, and then stored in another tank. The rich solution is recirculated to its reservoir.

The thermal coupling is observed during the discharging phase (Figure 2.13b): as is the case in the one-stage cycle, the working fluid is circulated from its reservoir to the evaporators 1 and 2 and the solution is circulated from its reservoir to the absorbers 1 and 2. The evaporator 2 is heated with an external source and the solution from the absorbers, enriched, and then returns to its reservoir. In addition, the two stages of the cycle exchange heat. Here, absorber 2 and evaporator 1 are thermally connected: the absorption heat of absorber 2 is used to heat and increase the temperature of evaporator 1. Thus, the operating pressure during stage 1 is higher than in stage 2. The steam from evaporator 1 is absorbed in absorber 1, and the absorption phenomenon is then obtained at a high temperature and can be used to produce DHW or the heating of interior spaces. This cycle therefore makes it possible to use the same external source temperature for charging as used in the one-stage cycle (Figure 2.14(a)) to obtain a higher temperature of heat produced during discharging (Figure 2.14(b)) and for the same conditions of external sources.

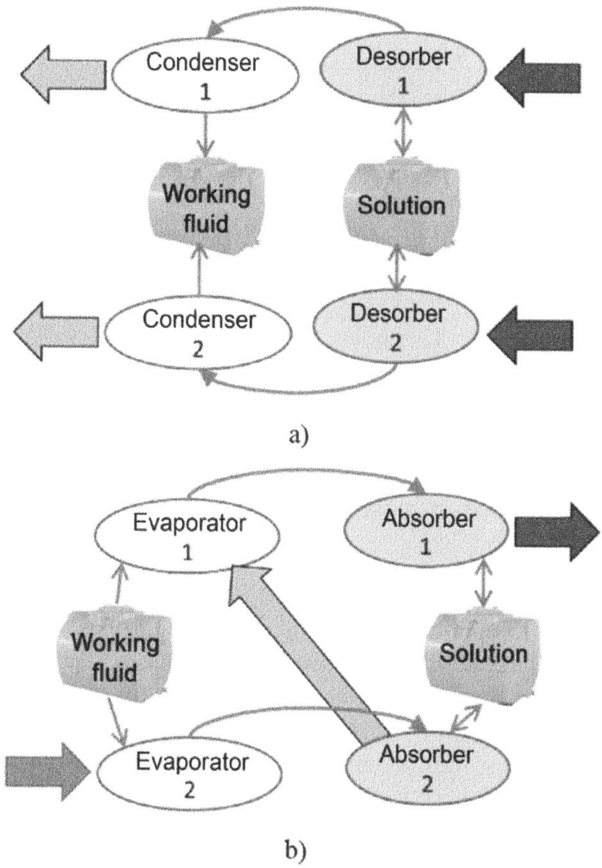

Figure 2.13. *Principle and operation of the two-stage absorption storage method: (a) charging phase; (b) discharging phase*

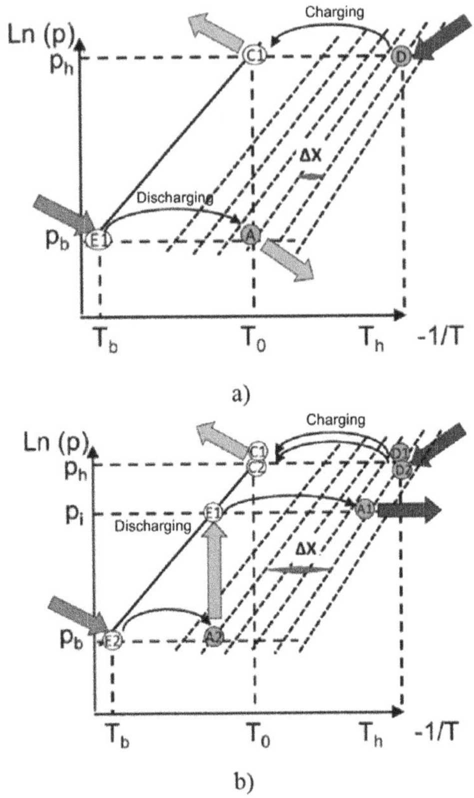

Figure 2.14. *Operating conditions of the two-stage absorption storage method with external heat exchange (A: absorber; D: desorber; E: evaporator; C: condenser; ΔX: acceptable difference in mass fraction for the cycle): (a) one-stage cycle; (b) two-stage cycle*

Observing how it operates during the discharging phase (Figure 2.14(b)), only the solution that circulates in absorber 1 produces heat that is beneficial to the user. The heat produced in absorber 2 is used for the internal operation of the cycle. Therefore, for the same amount of stored solution, the cycle theoretically produces half as much useful heat as the one-stage cycle. However, in the case where the external temperatures vary greatly between the charging phase and the discharging phase, this impact can be made up for or even exceeded by the fact that the mass fractions of the solution can vary more significantly in this two-stage cycle than in the one-stage cycle, as shown in Figure 2.14(b). In fact, in the discharging phase, the temperature of the evaporator determines the operating pressure in the system, and therefore the minimum temperature of the useful heat determines the minimum acceptable mass fraction of the solution. In the charging phase, it is the temperature in the condenser which sets the high operating pressure in the system necessary, while the temperature of the heat source establishes the maximum possible mass fraction of the solution. These two conditions define the range of the mass fraction of the cycle (represented by the ΔX in Figure 2.14). In the case of a two-stage cycle, the discharging phase is no longer a constraint, since the temperature of the external source only sets the pressure of stage 2 of the system, and as a result, the minimum mass fraction that would allow the production of useful heat by the system can be reduced (Figure 2.14(b)). Thus, by expanding the range of the mass fraction of the solution between the two operating phases, it is possible to increase the energy storage density of the process. In comparison with the improved cycle presented in section 2.4.1, here the issue is no longer that of increasing the high mass fraction of the cycle, but of making it possible to decrease the low mass fraction. The results of the calculations on theoretical cycles have shown that the energy storage densities can be multiplied by 7 for a two-stage cycle, compared to the simple conventional cycle for the sorption couple LiBr-H_2O under certain operating conditions (Xu and Wang 2019). Overall, this cycle is therefore more complex than the one-stage cycle, but under unfavorable external operating conditions (such as high differences in the external temperature between charging and discharging), it makes it possible to maintain a satisfactory discharging temperature with high energy densities.

2.4.2.2. *Two-stage cycle with solution exchanger and internal heat*

Another intensified cycle has also been proposed for storage. The development of this two-stage cycle also requires bringing together two similar stages operating in parallel: the system consists of two absorbers/desorbers (referred to as 1 and 2 in Figure 2.15) and two evapo-condensers (also referred to as 1 and 2).

This cycle was proposed by Weber and Dorer (2007). During the discharging phase (Figure 2.15(b)), this cycle works exactly like the previous two-stage cycle, with the thermal coupling of the two stages of the system. The vapor from evaporator 1 heated by absorber 2 is absorbed in absorber 1, and its absorption heat is obtained at a high temperature. This cycle therefore makes it possible to obtain a higher heat temperature produced during the discharging from the same external source temperature as in the one-stage cycle.

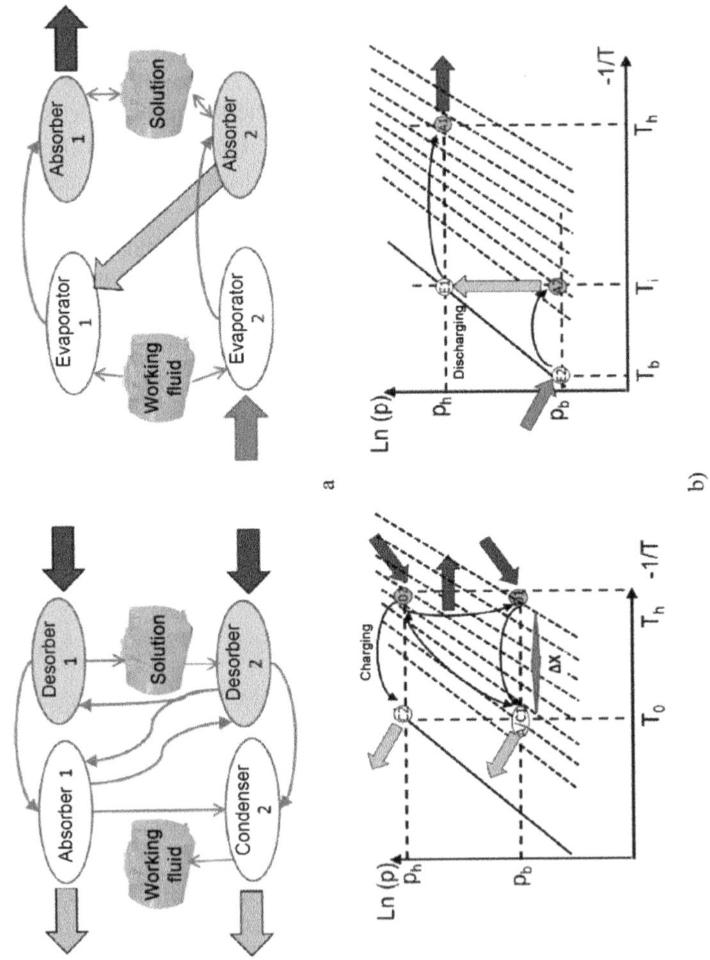

Figure 2.15. *Principle and operation of the two-stage absorption storage method with an exchange of the solution and internal heat: (a) charging phase; (b) discharge phase*

This cycle therefore has two advantages: it makes it possible to obtain a discharging temperature higher than the one-stage cycle, and it makes it possible to use a heat source at a more moderate temperature during charging than the one-stage cycle for the same enrichment of the solution. For example, in the case of the NaOH-H_2O sorption couple, the charging temperature required for the two-stage cycle could be reduced to 95°C instead of 150°C for the one-stage cycle, in order to obtain the same high mass fraction at the end of the charging, as reported by Weber and Dorer (2007). However, this cycle also has the disadvantage of being more complex: here, two cycle stages must be managed in parallel during the two operating phases. It also requires more active components (exchangers), which increases the cost and the size of the system. In addition, condenser 1 sees the solution circulate during the charging phase, and if salt is deposited on the walls, this could lead to impurities in the working fluid that evaporates in the same component (evaporator 1) in the discharging phase.

Finally, it can be observed that, for the previous two-stage cycle, the storage capacity of the ideal cycle can either be reduced or improved compared to the one-stage cycle, due to the discharging phase. In addition, during the charging phase in this case, heat must be transmitted to the two desorbers in order to reach the desired mass fraction, which also reduces the COP of the ideal two-stage cycle compared to the one-stage cycle. This two-stage cycle must therefore be reserved for cases where the heat source used in the charging phase is free heat, as is the case when industrial thermal heat waste or solar heat is recovered.

2.4.2.3. *Cold storage: two-stage, two-phase cycle*

The absorption process has the advantage of allowing heat storage, as presented in the previous cases, but also with the ability to store a chemical potential to produce cooling, as permitted by the AHP cycle shown at the beginning of this chapter. In this case, one potential improvement in the performance of the system is the development of a multistage two-phase cycle.

This cycle design is shown in Figure 2.16, showing the case of the two-phase series cycle.

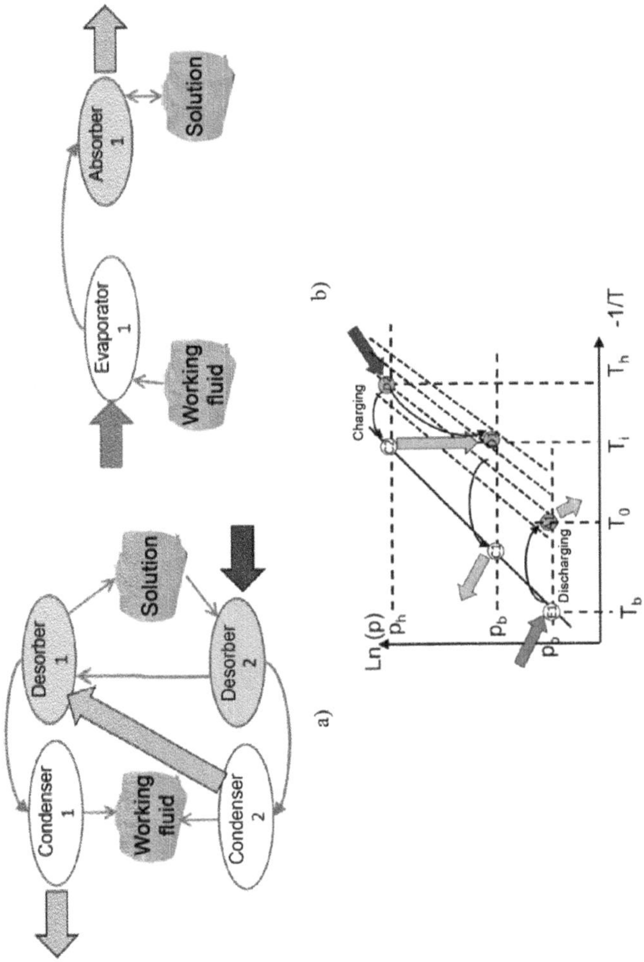

Figure 2.16. *Principle and operation of the two-phase cold storage method: (a) charging phase; (b) discharging phase; (c) Oldham diagram*

Here, the discharging phase remains the same as in the one-phase cycle. Here, the useful output is the cold produced at the evaporator at low temperatures, while the absorption heat is rejected at a temperature close to ambient temperature. The difference is observed in the charging phase, for which one desorber/condenser at two pressure levels is thermally coupled. What is of interest here is to desorb a greater quantity of working fluid from the same quantity of heat. As we can see (Figure 2.16(a)), the desorber 2 consumes some of the heat from the heat source. This leads to desorption of part of the working fluid, which condenses at high pressure and relatively high temperature in condenser 2. This condenser is in thermal contact with desorber 1, into which the solution with an intermediate mass fraction exiting from desorber 2 is transferred. Thus, the condensation heat of condenser 2 allows for the desorption of the working fluid in desorber 1 at an intermediate pressure (and therefore temperature) Ti. The working fluid that is desorbed in this way can condense at a temperature close to the ambient temperature in condenser 1. Thus, the internal heat transfer between condenser 2 and desorber 1 makes it possible to desorb a greater quantity of working fluid than in the simple cycle from the same quantity of heat taken from the heat source at Th (Figure 2.16(c)). This cycle was studied in particular by De and Ganguly (2020) for the couple LiBr–H_2O, in a case where two heat exchangers are inserted between the solution tubes without tanks. In this case, an increase in the COP of the system of almost 80% for the same cold temperature level was observed. On the other hand, the energy efficiency of this two-stage system is only about 16% higher than that of the one-stage system. In fact, in this cycle, the temperature needed for the heat that is supplied to desorber 2 is higher than in the case of the one-stage cycle, to allow a heat transfer at an intermediate temperature.

2.5. Integration of absorption storage systems: case studies

As seen, the processes of heat storage by absorption have not yet reached technological and industrial maturity. Currently, several prototypes have been developed and tested in research laboratories, and the feasibility and interest in the concept have been demonstrated. Intensified cycles have also been developed. Nevertheless, these storage systems will require significant investment costs due to both the sorption materials and the specific reactors to be implemented, or to their operating under a vacuum, which implies significant technological constraints. Nevertheless, analyses have been developed through modeling to propose possibilities for integrating these systems in particular contexts and to study their technical and economic relevance.

2.5.1. *Integration of absorption heat storage systems in solar buildings*

Sorption heat storage processes make it possible to operate using cycles of duration adjustable to use needs, due to the fact that the heat is stored in the form of a chemical potential. As a result, one application of this type of storage is the inter-seasonal storage of heat from thermal solar collectors in summer for the heating of buildings during the following winter, as shown in Figure 2.17. The use of the excess heat produced in the summer makes it possible for the solar collectors to be taken advantage of over a longer period of the year and to increase the solar coverage rate of the overall heating and DHW needs of a building.

For this absorption process, the operating conditions of the system, and in particular the temperature to be reached during the production of heat by the sorption system, will have a significant influence on the performance of the process. For example, Figure 2.18 (N'Tsoukpoe 2012) shows the size of the storage as a function of the maximum mass ratio of crystals in the solution tank for temperatures of the heated floor of the building (Tgi Figure 2.18) of 30°C and 33°C. These volumes and masses are calculated in the case of storage with the couple LiBr–H_2O to cover the annual heating needs of a single family house with a very efficient floor area of 120 m^2, storing 1,800 kWh/year (excluding DHW needs) in the climate of Chambéry, in France.

In this modeling, it is assumed that there is a direct coupling of the solar collectors with the sorption storage, or a direct coupling of the storage with a heating floor, depending on the availability or need for heat. This model takes into account the dynamic behavior of the system and its operation throughout the year. This analysis highlights the cycles that occur primarily in the transition seasons (spring and autumn), which allow the system to be partially charged on sunny days, and for the storage to be partially discharged on cooler days, based on the needs of the building. Overall, by taking all the partial charging and discharging over a year into account, the storage achieves between 1.5 and 2 charging/discharging cycles per year.

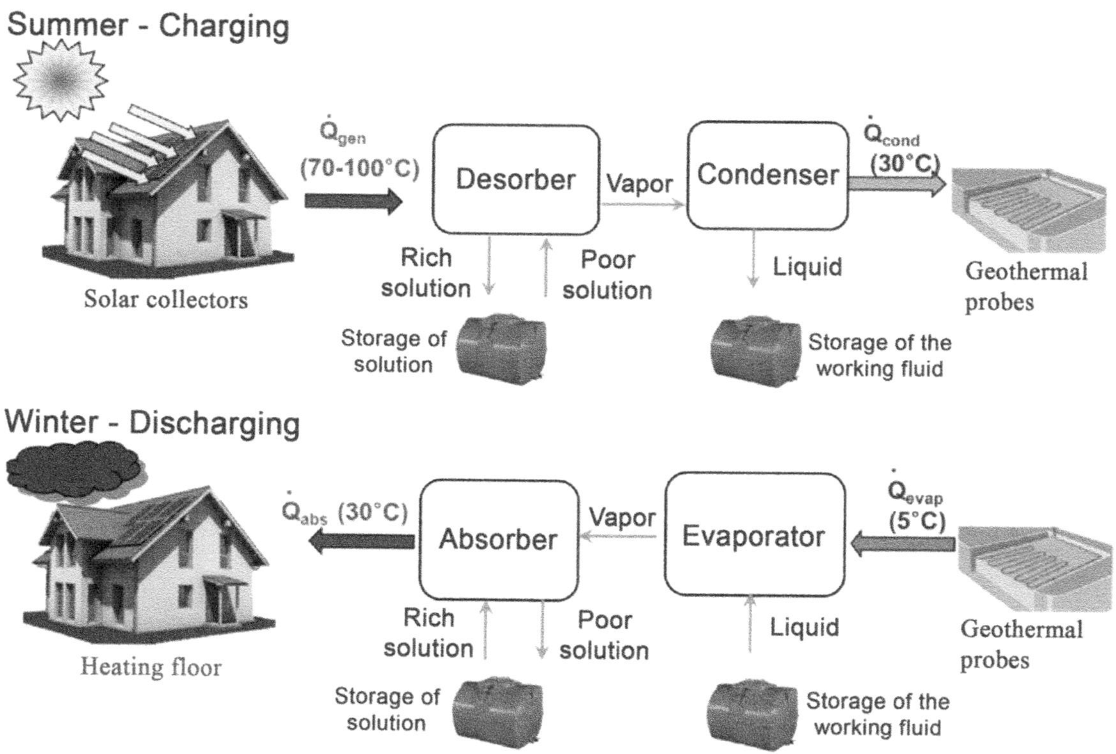

Figure 2.17. *Operating principle of the process of interseasonal storage of solar heat by absorption integrated into a building*

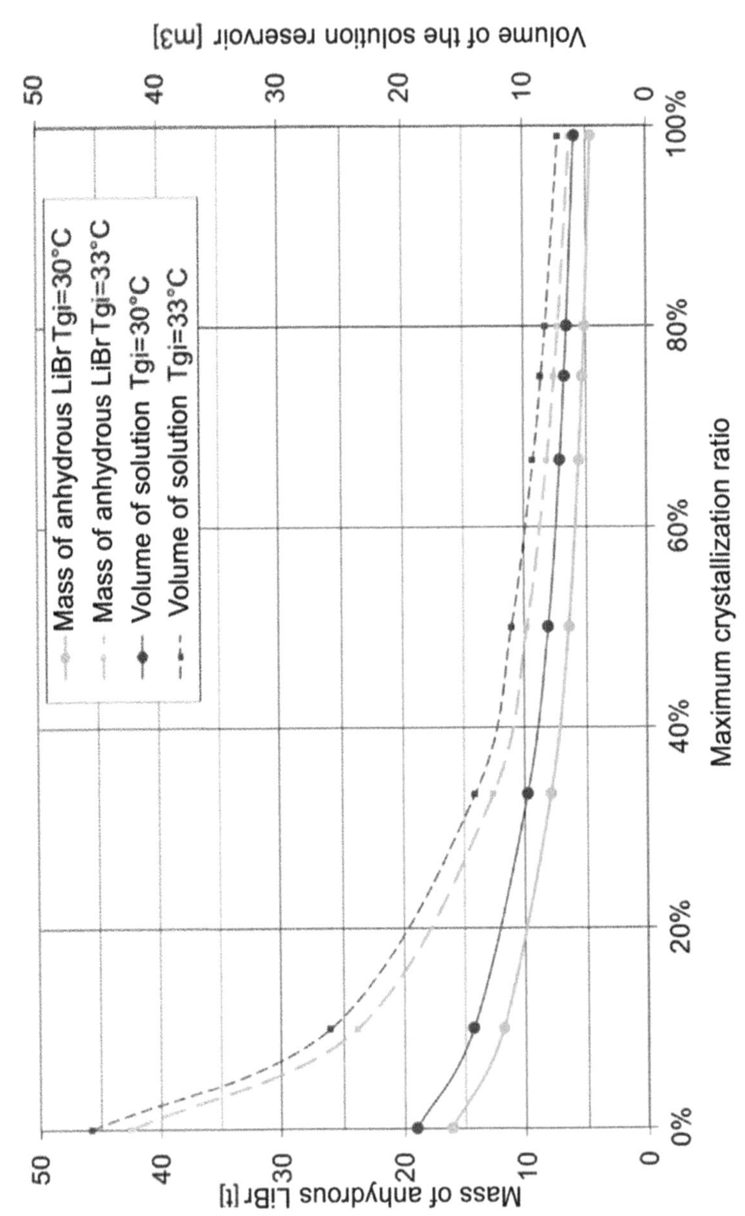

Figure 2.18. *Effect of the crystallization ratio and the average temperature of the discharging heat transfer fluid on the necessary volume and mass for the storage of 1,800 kWh (N'Tsoukpoe 2012)*

Aside from the interest in crystallization, which has already been examined previously (section 2.4.1), Figure 2.18 highlights the influence of the heating temperature to be reached in the energy density from storage. This is sometimes divided by more than 2 when the temperature to be reached increases by only 3°C (though this ratio decreases with the crystallization ratio). Indeed, to obtain significant absorption temperatures, the minimum mass fraction of the solution during the cycle must be sufficiently high (see Figure 2.11, point 2d). Consequently, the storage density decreases when the minimum absorption temperature increases. The process is therefore better adapted to high-performance buildings equipped with low-temperature heat emitters, such as heating floors. In addition, the production of DHW using this system alone is not feasible. However, it can contribute to the preheating of the DHW if it is assisted by an additional heating device. In addition, the storage volumes must be increased in order to cover all annual heating needs of this individual house (on the order of 10 m^3 for a crystallization ratio of 50%); in this case, the necessary surface area of the solar collector of about 13 m^2 (N'Tsoukpoe 2012) is quite high, and difficult to conceive of in future applications. Thus, combining these systems with an additional heating unit is preferable.

2.5.2. *Integration of absorption heat storage systems into a decentralized microgrid*

Another analysis of the integration of heat storage by absorption was conducted using the case of a decentralized energy microgrid (Wang et al. 2019). This application does not involve storing heat for heating buildings, but instead seeks to cool them, as mentioned previously (Figure 2.19). One case study evaluated the performance of a decentralized microgrid in providing both electricity and cooling for buildings with a floor area of approximately 250,000 m^2 on a university campus, in the climate of Hong Kong. In this example, an electrical production plant provides for users' electricity needs and the electrical consumption of air conditioners (a vapor compression heat pump). In the case where the storage is integrated into the microgrid, batteries are added for electrical storage, and thermal storage by absorption makes it possible to recover the heat discharged by the electric power plant (Figure 2.19) for the production of cold temperatures necessary for air conditioning. The modeling of the operation of this system is based on a number of simplifications, such as an energy storage density for absorption with the LiBr–H_2O couple fixed at 151.1 kWh/m^3. The proposed absorption storage therefore has a much higher energy density than the ice water storage that is often used together with electric air conditioners using vapor compression (on the order of 10 kWh/m^3).

The scaling of the storage system called for the addition of 63 tanks of 10 m^3 for storing the solution. This also makes it possible to reduce the number of air conditioning units using vapor compression from 10 to 6, and thus the electrical power used from 20.5 to 12.3 MW. The grid that uses storage (electrical storage and absorption storage) has better performance than the grid without energy storage. The electrical power for the air conditioners during peak hours is reduced by 17.5%, since part of the cooling is produced by the absorption system. In addition, compared to the grid without storage, the average usage rate of the decentralized thermal power plant increased from 80% to 92.9%, which led to higher efficiency in the electric cycle (less the operation at a partial load) and a decrease in aging (extended on/off cycles). The operating costs can therefore be reduced by 12.9%, and the overall primary energy consumption of the microgrid also decreases. Finally, the peak electricity consumption, which had to be imported from the electricity grid in the case where storage is not used, can be fully managed internally from this decentralized microgrid, which becomes self-sufficient in its electricity use.

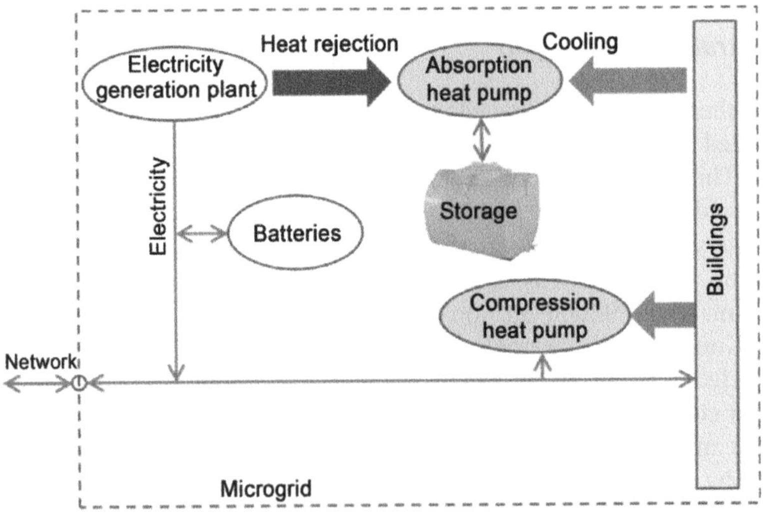

Figure 2.19. *Integration of heat storage by absorption into a decentralized microgrid (Wang et al. 2019)*

In this case, storage by absorption thus seems to be relevant in terms of both energy use and economic concerns. This is particularly due to the fact that the number of annual charging/discharging cycles is much greater than in the case of inter-seasonal heat storage; here it is one cycle per day, versus only one per year (or two cycles per year depending on the partial cycles in the off-season) in the case of summer-winter storage. This application, with its shorter duration cycles, makes it easier to justify a significant initial investment. Therefore, the best technical-economic compromise for the use of these absorption storage systems must be studied further according to the economic conditions and the constraints of the resource and the energy needs.

2.6. Conclusion

As seen, the technology of heat storage by absorption opens up many possibilities, both in terms of storage duration and in terms of energy storage density. For instance, noteworthy storage densities of 54 kWh/m^3 (relative to the volume of the prototype) or 110 kWh/m^3 (relative to the volume of the solution and water) have been obtained experimentally on various prototypes, and much higher densities have been determined through numerical simulations on optimized cycles under favorable conditions. Nevertheless, no commercial system is available on the international market as of this time. Several research teams have focused on the various scientific issues raised by these processes, whether in terms of the sorption materials (Chapter 1), the exchangers/reactors to be used, the various possible optimized cycles or the technological problems caused by the operation of the system under vacuum. In general, it is quite clear that a good reactor design is critically important for obtaining the desired heat and mass transfer processes. A few weak points were noted: some prototypes tested had problems with the wettability of the exchange surfaces and the presence of non-condensable gases in the reactors, which had a strong negative impact on their performance. This problem of the non-condensable introduction into the system through leaks arises all the more frequently since the system must be able to operate over very long periods, lasting several decades, in order to be able to justify a significant initial financial investment. The crystallization of the solution, which should make it possible to further intensify the energy density of the system, also creates risks in terms of the circulation of fluids in certain configurations of the system. Furthermore, the optimization of the cycle itself would make it possible to improve the efficiency of the systems.

Thus, additional studies must still be carried out so that thermal storage by absorption can be brought to its next level of development. In particular, these studies will have to examine the following areas:

– the development of new exchangers/reactors. The exchangers used so far have faced several problems, in particular related to the wetting of the exchange surfaces. In addition, for storage using crystallization, a new reactor design adapted to the specific risks arising from the crystallization of the solution would be very useful to obtain a very high storage density;

– the development of new advanced cycles of storage by absorption. The simple single-stage storage cycle is still under development, and some of the problems identified for this cycle could potentially be solved and thus allow for better performance by using intensified cycles;

– the development of new materials, as has already been expanded on in Chapter 1. The materials to be used should combine good thermodynamic characteristics, good environmental safety and low costs. Currently, no tested couple combines all these qualities, so new sorption couples should be studied in the future.

2.7. References

ADEME (2017). La chaleur fatale industrielle [Online]. Available at: https://www.ademe.fr/sites/default/files/assets/documents/chaleur_fatale-8821-2018-06pdf.pdf.

Altamirano, A., Le Pierrès, N., Stutz, B. (2019). Review of small-capacity single-stage continuous absorption systems operating on binary working fluids for cooling. Theoretical, experimental and commercial cycles. *International Journal of Refrigeration*, 106, 350–373.

Altamirano, A., Stutz, B., Le Pierrès, N. (2020). Review on small capacity single-stage continuous absorption systems operating on binary working fluids for cooling: Compact heat exchanger technologies. *International Journal of Refrigeration*, 114, 11–147.

Bales, C. and Nordlander, S. (2005). TCA evaluation – Lab measurements, modelling and system simulations. Report, SERC, Högskolan Dalarna, Borlänge.

Daguenet-Frick, X., Gantenbein, P., Müller, J., Fumey, B., Weber, R. (2017). Seasonal thermochemical energy storage: Comparison of the experimental results with the modeling of the falling film tube bundle heat and mass exchanger unit. *Renewable Energy*, 110, 162–173.

Descombes, G. and Desmet, B. (2023). Energy conversion: Thermodynamic basics. In *Thermodynamics of Heat Engines*, Desmet, B. (ed.). ISTE Ltd, London, and John Wiley & Sons, New York.

Gao, J.T., Xu, Z.Y., Wang, R.Z. (2020). Experimental study on a double-stage absorption solar thermal storage system with enhanced energy storage density. *Applied Energy*, 262, 114476.

Herold, K.E., Radermacher, R., Klein, S.A. (2016). *Absorption Chillers and Heat Pumps*. CRC Press, Boca Raton.

Hiu, L. (2010). Stockage inter-saisonnier d'énergie solaire pour l'habitat par absorption. PhD Thesis, Université de Grenoble, Grenoble [Online]. Available at: https://tel.archives-ouvertes.fr/tel-00574180.

Huaylla Roque, F. (2017). Stockage d'énergie solaire par absorption : expérimentation et modélisation du procédé. PhD Thesis, Université Grenoble Alpes, Grenoble [Online]. Available at: https://tel.archives-ouvertes.fr/tel-02050102.

Le Pierrès, N., Huaylla, F., Stutz, B., Perraud, J. (2017). Long term solar heat storage process by absorption with the pairing KCOOH–H_2O: Experimental investigation. *Energy*, 141, 1313–1323.

Minvielle, Z., Caney, N., Clément, P., Bandelier, P., Marty, P., Gruss, J.A., Roussel, C. (2014). Échangeurs de chaleur – Description. *Techniques de l'Ingénieur*, BE9516 v1.

Nikbakhti, R., Wang, X., Hussein, A.K., Iranmanesh, A. (2020). Absorption cooling systems – Review of various techniques for energy performance enhancement. *Alexandria Engineering Journal*, 59(2), 707–738.

N'Tsoukpoe, K.E. (2012). Étude du stockage à long terme de l'énergie solaire thermique par procédé d'absorption LiBr–H_2O pour le chauffage de l'habitat. PhD Thesis, Université de Grenoble, Grenoble [Online]. Available at: https://tel.archives-ouvertes.fr/tel-00710249.

N'Tsoukpoe, K.E., Le Pierrès, N., Luo, L. (2013). Experimentation of a LiBr–H_2O absorption heat storage process: Prototype design and first results. *Energy*, 53, 179–198.

N'Tsoukpoe, K.E., Perier-Muzet, M., Le Pierrès, N., Luo, L., Mangin, D. (2014). Thermodynamic study of a LiBr–H_2O absorption process with crystallisation of the solution for solar thermal storage. *Solar Energy*, 104, 2–15.

Perier-Muzet, M. and Le Pierrès, N. (2016). Modeling and analysis of energetic and exergetic efficiencies of a LiBr/H_20 absorption heat storage system for solar space heating in buildings. *Energy Efficiency*, 9, 281–299.

Quinnell, J.A. and Davidson, J.H. (2014). Heat and mass transfer during heating of a hybrid absorption/sensible storage tank. *Solar Energy*, 104, 19–28.

RTE (2020). Réduction des émissions de CO_2, impact sur le système électrique : quelle contribution du chauffage dans les bâtiments à l'horizon 2035 ? [Online]. Available at: https://assets.rte-france.com/prod/public/2021-01/Rapport%20chauffage_RTE_Ademe.pdf.

Srikhirin, P., Aphornratana, S., Chungpaibulpatana, S. (2001). A review of absorption refrigeration technologies. *Renewable and Sustainable Energy Reviews*, 5(4), 343–372.

Stutz, B., Altamirano, A., Stepec, M., Le Pierrès, N. (2016). Evaporateurs à films ruisselants. In *Actes du Congrès Français de Thermique*. SFT, Toulouse.

Stutz, B., Le Pierrès, N., Kuznik, F., Johannes, K., Palomo Del Barrio, E., Bédécarrats, J.P., Gibout, S., Marty, P., Zalewski, L., Soto, J. et al. (2017). Storage of thermal solar energy. *Comptes Rendus Physique*, 18(7/8), 401–414.

Wang, L., Xiao, F., Cui, B., Hu, M., Lu, T. (2019). Performance analysis of absorption thermal energy storage for distributed energy systems. *Energy Procedia*, 158, 3152–3157.

Weber, R. and Dorer, V. (2008). Long-term heat storage with NaOH. *Vacuum*, 82, 708–716.

Weiss, W. and Spörk-Dür, M. (2019). Solar heat worldwide, global market. Development and trends in 2018. Detailed market figures 2017. Report, IEA Solar Heating & Cooling Program.

Xu, Z.Y. and Wang, R.Z. (2019). Absorption seasonal thermal storage cycle with high energy storage density through multi-stage output. *Energy*, 167, 1086–1096.

Zhang, X., Li, M., Shi, W., Wang, B., Li, X. (2014). Experimental investigation on charging and discharging performance of absorption thermal energy storage system. *Energy Conversion and Management*, 85, 425–434.

Zhao, Y.J., Wang, R.Z., Li, T.X., Nomura, Y. (2016). Investigation of a 10 kWh sorption heat storage device for effective utilization of low-grade thermal energy. *Energy*, 113, 739–747.

3

Heat Storage Using Adsorption Processes

Larysa RATEL[1], Kevyn JOHANNES[2] and Frédéric KUZNIK[1]
[1] *CETHIL, CNRS, INSA Lyon, Villeurbanne, France*
[2] *CETHIL, CNRS, Université Claude Bernard Lyon 1, Villeurbanne, France*

3.1. Introduction

The storage of heat is a method that can be used to resolve some of the issues involving discrepancies between thermal demand and provision in terms of both time and power. Currently, there are several types of known heat storage, such as sensitive heat storage, latent heat storage, sorption heat storage and chemical heat storage. In this chapter, we will focus on the storage of heat by sorption, and more precisely by *adsorption*.

Several authors have attempted to clarify the definition of sorption storage systems (Srivastava and Eames 1998; Hauer 2007a; N'Tsoukpoe et al. 2009; Abedin 2011; Kousksou et al. 2014; Kuznik and Johannes 2020). Sorption is defined as a phenomenon involving the fixation or capture of a gas (sorbate) by a body in a condensed state, such as a solid or a liquid (sorbent) (Kousksou et al. 2014; Sun et al. 2016). *Adsorption*, in the context of gas storage, is a surface phenomenon in which a gas becomes affixed to the surface of a solid or a porous material (Sun et al. 2016; Tulus et al. 2016). The mechanism of the adsorption heat storage process is based on a reversible transfer, which can be given as:

$$S \cdot nG + \mathrm{Q} \leftrightarrows S + nG \text{ (I)}$$

Heat and Cold Storage 2,
coordinated by Nolwenn LE PIERRÈS and Lingai LUO. © ISTE Ltd 2024.

where S is the sorption material and G is the sorbate, SG is the sorption pairing and Q is the adsorption energy. In adsorption by heat storage, water vapor is generally used, as described in Chapter 1.

3.2. Overview of heat storage by adsorption

3.2.1. *Principle of the operation of adsorption materials*

The configuration of a solid/gas adsorption system for the production of heat consists of an adsorbent heat exchanger operating in alternating connection with an evaporator or a condenser, which depends on the stage of the cycle, as well as a set of valves (Figure 3.1(a)). The adsorbent tank acts as an adsorber and a desorber through the discontinuous circulation of adsorbate in the system. A schematic of the complete adsorption cycle is given in the Clausius Clapeyron diagram (($\ln\left(\frac{p}{p_0}\right)$ depending on $-\frac{1}{T}$) in Figure 3.1(b)), which consists of four steps: isobaric adsorption ($1 \rightarrow 2$), isosteric heating ($2 \rightarrow 3$), isobaric desorption ($3 \rightarrow 4$) and isosteric cooling ($4 \rightarrow 1$) (Pinheiro et al. 2015, 2020; Elsheniti et al. 2018; Kuznik et al. 2018; Xu et al. 2018).

Each cycle develops according to the following scenarios. During isobaric adsorption ($1 \rightarrow 2$), valve 2 is open and the adsorbent tank is connected with the evaporator, which is at a temperature lower than the thermodynamic adsorbtion equilibrium temperature. The steam produced in the evaporator is adsorbed by the adsorbent and the heat is produced by the system, so as to then be used for heating the surrounding ambient space. During isosteric heating ($2 \rightarrow 3$), the two valves are closed and the adsorbent tank is heated by an external source. The adsorbate is partially desorbed, leading to an increase in the pressure in the system from P_E to P_C. Once the pressure has reached that of the condenser, valve 1 is opened and the isobaric desorption ($3 \rightarrow 4$) begins. The temperature of the adsorbent bed increases and the adsorbate is released and condensed in the condenser. There is a latent amount of heat from the phase change of the adsorbate from the gaseous state to the liquid state, which can be recovered in the condenser and used for heating. Finally, the two valves are closed, the temperature and the pressure in the tank containing the adsorbent decrease isosterically, and once the pressure reaches the pressure P_E, a new cycle can be initiated by connecting the adsorbent tank to the evaporator.

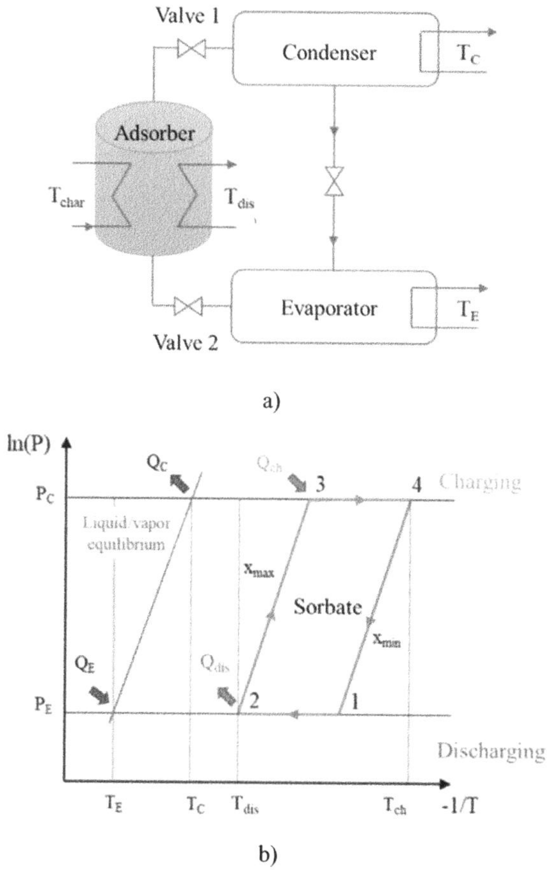

Figure 3.1. *Configuration of a solid/gas adsorption system and the operating principles of an adsorption cycle*

3.2.2. *Classification of systems*

The existing configurations of systems for heat storage using adsorption can be divided according to a comprehensive vision of the system. These systems can be defined using two primary criteria: whether they are closed or open system, and whether the reactor is integrated into or separated from the storage tank. In closed systems (which were described briefly in the previous section), the mass transfer occurs in a closed loop between the adsorber, a condenser and an evaporator. In open systems, the mass exchange occurs with the area outside the system in order to supply the adsorbate to the reactor.

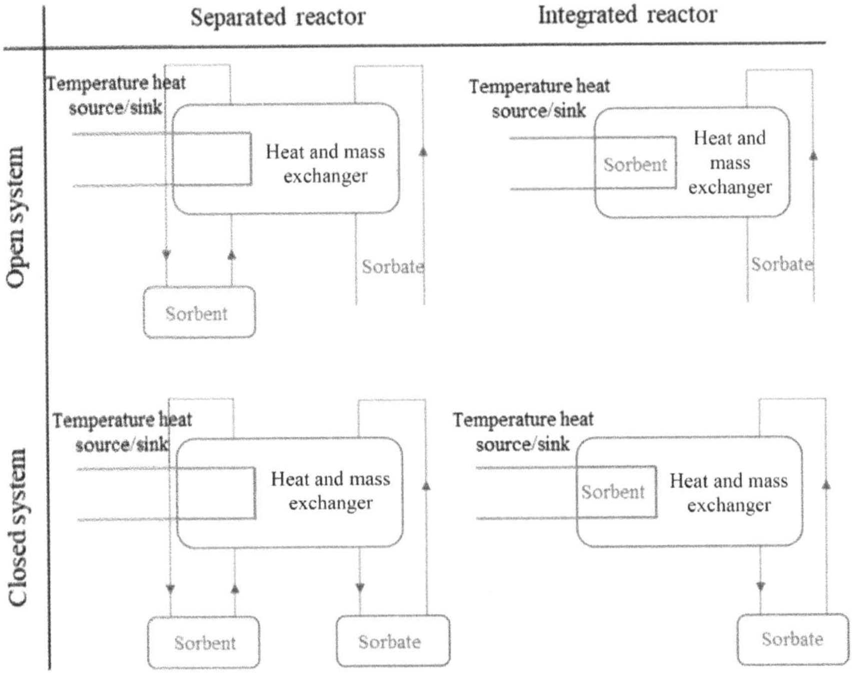

Figure 3.2. *Classification of heat storage systems using adsorption*

The four sorption heat storage processes are shown in Figure 3.2.

3.2.2.1. *Storage systems: open or closed*

In open systems, an exchange of mass and heat with the outside environment occurs (Figure 3.3(a)). This system allows for operation at pressure levels that are typically ambient pressure, because the adsorbate used is taken and returned to the atmosphere (Weber et al. 2016). Humid air also acts as a heat transfer fluid and adsorbate source (water vapor) (Van Alebeek et al. 2018). The heat released by this step can be used in such ways as heating people's homes. The configuration of such a system is relatively simple, consisting of a reactor, which contains the adsorbent, and a fan, which allows the circulation of a stream of moist air in the reactor. This system can be equipped with a humidifier in order to increase the moisture content of the air flow passing through the reactor (Tatsidjodoung et al. 2016). During the charging phase of the system, the ambient air is heated by a heat source and circulates through the reactor containing the adsorbent. The heat is transferred to the sorption material, which causes the adsorbed water to begin the process of desorption. The water vapor generated and thus transported through the flow of air is discharged back into the outside environment. During the discharge phase, cold moist air is transported into the reactor containing the adsorbent. The water vapor contained in this humid air is adsorbed by the material, which allows for heat to be released in the reactor, and the dry and hot air produced is used for purposes such as heating the habitat.

The major advantages of an open system are the simplicity of the configuration, and therefore the low costs of investment in the system's manufacture and operation, since the operation is done at ambient pressure and the heat transfer can be controlled. The biggest disadvantages involve the availability of water vapor at the time of discharge in order to ensure a correct sorption speed. The use of a fan is essential in order to ensure that the flow of air circulates in the reactor. By using a fan and a humidifier, the electrical energy consumed by the system is increased. The choice of adsorption material must be done carefully, ensuring that hazardous materials or materials containing toxic components are not used, as described in Chapter 1. In addition, due to mass exchanges with the environment, the potential degradation of adsorbent materials due to impurities and the contamination of the air flow by fine particles must be addressed (Bennici et al. 2020).

In closed systems (Figure 3.3(b)), no mass exchange with the environment occurs, and only energy exchanges are possible (Zondag 2015). This system operates under a partial vacuum (at the phase change pressure of the adsorbate, as described in Chapter 2 in the case of absorption) in order to make sure the adsorbate is transferred between the reactor and the tank containing the adsorbate. The principles of this operation are shown in Figure 3.3(B). In general, this type of system is composed of two separate tanks: the reactor, containing the reactive material, and the evaporator/condenser, which supplies and recovers the adsorbate. During the system charging stage, the heat supplied by an external heat source is transferred to the reactor, which causes desorption. The adsorbate passes into the gaseous state and is sent to the condenser/evaporator, where it is condensed to the liquid state. At the end of this process, the reactor and the condenser are separated, and the materials are kept separate until wintertime. During the discharge of the system, the reactor is connected to the evaporator, where the liquid adsorbate is evaporated at a constant temperature using an external energy source. The heat produced by sorption is transmitted to nearby living spaces. In the summer, the evaporation process can be used to fulfill needs for cooling.

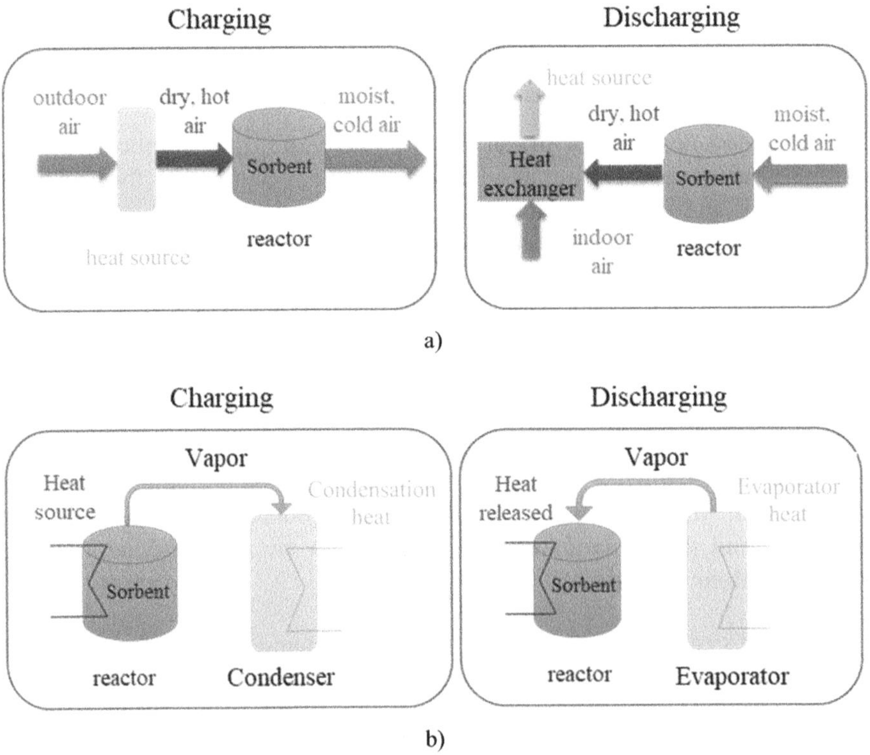

Figure 3.3. *Concepts of the operation of the (a) open and (b) closed heat storage. For a color version of this figure, see www.iste.co.uk/lepierres/heat2.zip*

Closed systems have several advantages for the storage of waste: they do not require the use of a fan, since the adsorbate is driven by the pressure difference between the components of the system and the transporting of the flow is faster with respect to open systems. The system can also operate using adsorbates that cannot be released into the environment such as ammonia; the combined use is conceivable in sorption heat pump for heating and cooling (Scapino et al. 2017). On the other hand, the major downsides of closed systems are the following: the need to install an evaporator/condenser, as well as additional devices to maintain the system phase change pressure of the adsorbate, which makes this system bulkier, more complex and therefore more expensive (Zhang and Wang 2020). The heat transfer is a limiting step (Michel et al. 2014), where a given amount of energy is to be supplied to the evaporator during the discharge process, and the possible condensation of the liquid within the system can block the flow of the adsorbate to the condenser.

A study carried out on the comparison of these two systems (Abedin and Rosen 2012) showed that energy efficiency and overall exergetic efficiency are better for open systems (69% and 23%, respectively) than for closed systems (50% and 9%, respectively). However, closed systems potentially require higher desorption temperatures, but also give off higher outlet temperatures than open systems, which is important when a high outlet temperature and a maximum discharge state need to be achieved. Despite this, the losses for both systems remain quite significant, and improving their efficiency is essential (Fumey et al. 2019).

3.2.2.2. *Type of reactor: separate or integrated*

Sorption thermal energy storage reactors can be divided into two main categories: separate or integrated reactors (Figure 3.4). In a separate reactor (Figure 3.2), the adsorbent reservoir and the mass and heat exchanger are located in different enclosures. The reactants are transported to the reactor where the reaction takes place, then returned to the storage tanks. In this type of storage, it is necessary to use separate storage tanks, as well as to provide means for transporting the reagents (Solé et al. 2015; Zondag 2015). Because it is possible to continuously transport the adsorbent and ensure the supply of fresh adsorbate, steady-state operation can be achieved. Therefore, continuous desorption and sorption can be maintained with constant temperature and power level settings. Another advantage of this type of system is that the sorption takes place on a fraction of the total amount of the adsorption material, and therefore a smaller volume of reactor can be used, which leads to low pressure drops, as well as to a less complex process control (Mette et al. 2013).

The integrated reactor contains all of the active material. Consequently, the solid material does not need to be transported, and therefore the electrical consumption of the transport is reduced, and no separate storage tank is necessary (Solé et al. 2015). The control of the sorption/desorption process through the reactive bed is very important and very complex (Fumey et al. 2019). This introduces certain parameters into the analysis, such as time and space, giving it a dynamic nature; and the various parameters, such as the transporting of masses, thermal conductivity, volume change, and dispersion at the scale of the bed, must all be considered. The disadvantage of this type of system is the need for a large reactor, and the management of the adsorption/desorption conditions. Extensive pressure losses across the reactive bed make it necessary to use electric fans, which in turn leads to an increase in energy consumption, and consequently, a decrease in energy storage efficiency.

The modular reactor can be considered as a subcategory of integrated or separate reactors (Mauran et al. 2008; Michel et al. 2012; Yu et al. 2013). The active material is stored in several sorption modules, the number of modules used is defined and their volume is optimized so as to achieve the desired storage capacity and release power. The use of modular reactors offers certain advantages. For example, an expansion or reduction of the systems requires low levels of technical effort, and the modules can be produced by standardized industrial procedures, allowing for competitive prices.

3.2.2.3. *Types of particle beds in the reactor*

A different classification system may be used, based on the type of reactor. First, it is important to note the distinction between batch reactors and continuous reactors. Batch reactors are often characterized by the addition of reactants into the reactor at the start of the reaction and a heating or cooling system that allows for heat transfers. The mass transfers are limited to the discharge of steam from the top of the tank. No air flow is intended to pass through the reactor. Batch reactors operate in non-stationary conditions. These are generally used in chemistry-related operation, from laboratory to industrial scales.

Continuous reactors are best suited for heat and mass transfer applications. Continuous reactors are designed to operate in stationary, tubular and agitated tanks. There are many types of reactors of different shapes and characteristics that have been studied for the purpose of solving the most significant problems, such as reducing heat losses, improving mass and heat transfers, and controlling the thermal capacity of heat. We can identify four types of reactors that are potentially of interest for sorption thermal storage systems: fixed bed reactors, fluidized bed reactors, screw reactors and gravity-assisted mass flow reactors (Zondag et al. 2009; Mette et al. 2012; Lichtarowicz 2013; Yu et al. 2013).

Fixed bed reactors are commonly used in a wide range of applications. When storing heat by adsorption, the solid reagent remains in the tank in the static state, while the fluid passes through the bed (Figure 3.4(a)).

In general, particles from a few millimeters up to a few centimeters in size are used. The fluid flows into the empty spaces of the bed with a flow rate between 0.1 and 1 $m.s^{-1}$ to ensure good mass and heat transfers between the fluid and solid phases, and to avoid excessive pressure loss (Lin et al. 2021).

The main advantage of fixed-bed reactors is the stratification of the reactor (Figure 3.4(b)). Using baffles can improve the passage of fluids and therefore the transfers of heat and mass. The heat is either transported by the inlet fluid or transferred to the reactor through a heat exchanger. In this case, the use of fins makes it possible to ensure effective conduction within the adsorbent bed in order to transport the heat to or from the heat-carrying fluid, as illustrated in Figure 3.5 (Makni 2012). The energy content of the thermal storage system can also be improved through the stratification of the reactor (Rosen 2001). Guilleminot et al. (1993) have proposed an alternative solution to fins for the improvement of heat transfers. They show that it is possible to increase the thermal conductivity of the bed by about 100 by replacing the pure adsorbent beads with a mixture containing a metal foam in addition to zeolite powder. This can reduce the size of the heat exchanger significantly and therefore lead to a more compact system. The advantage of fixed-bed reactors is their simple modeling and the ease of implementation, while the disadvantages are the relatively low mass and heat transfers as well as the pressure losses (Solé et al. 2015).

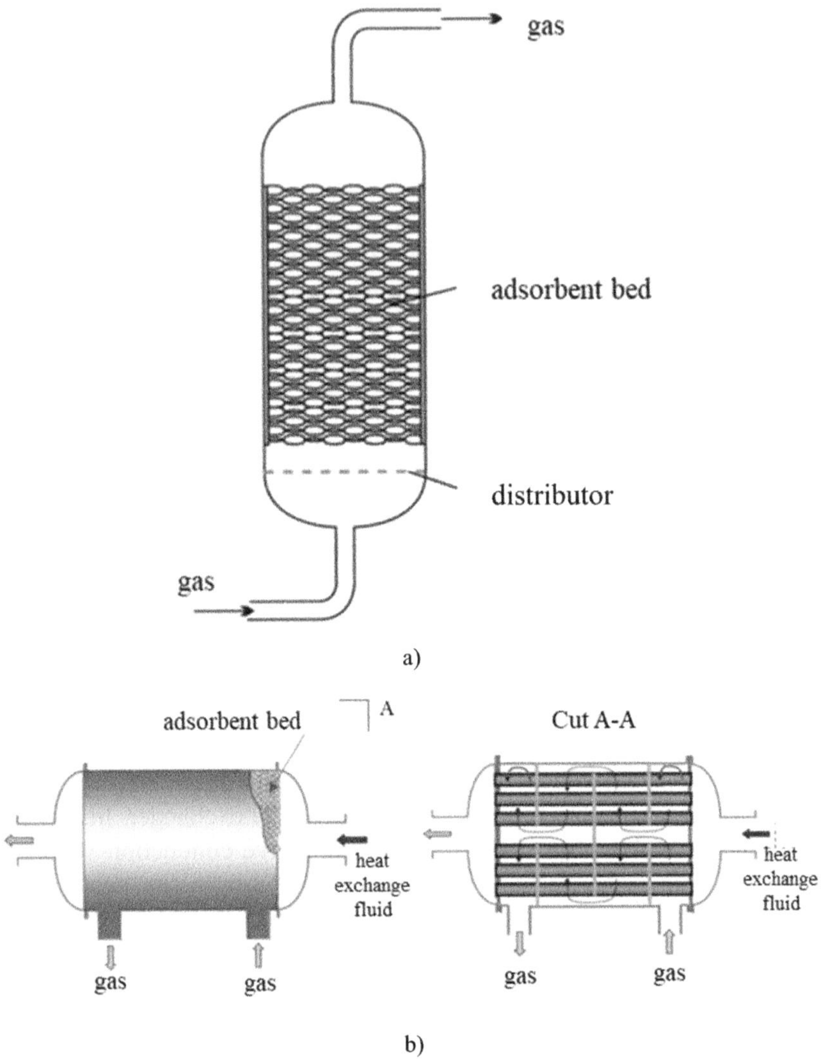

Figure 3.4. *Two examples of fixed bed reactors: (a) a simple reactor; (b) a reactor with an integrated heat exchanger and baffles (Lichtarowicz 2013)*

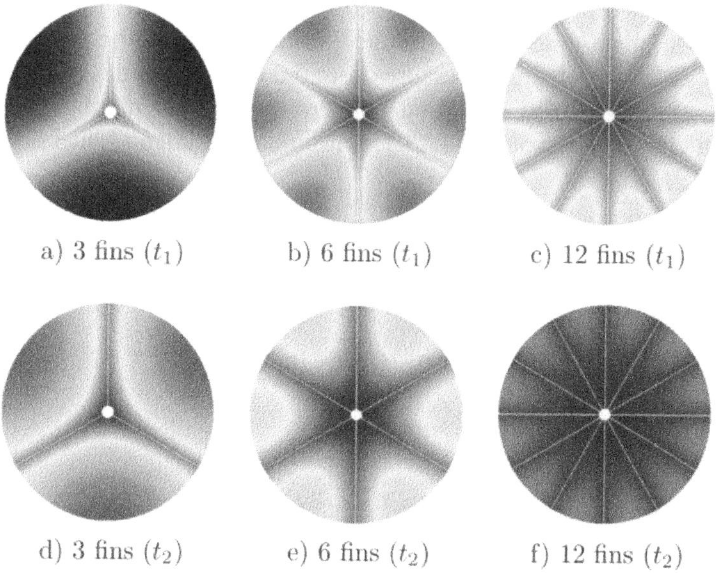

Figure 3.5. *Effect of the number of fins on the temperature profile in the adsorbent bed of a closed adsorption heat pump, with a heat transfer fluid circulating in a pipe integrated into the center of the cylindrical tank (Makni 2012). For a color version of this figure, see www.iste.co.uk/lepierres/heat2.zip*

In fluidized bed reactors, the solids are in the form of fine particles suspended by the circulation of the fluid. The input fluid flows from the bottom toward the top and applies an upward force that compensates for the force of gravity (Figure 3.6).

The movement of the particles creates a very good level of mixing, a uniform temperature distribution in the bed, and improves heat transfers to the substrate inside the bed, which can be advantageous for the application of heat storage (Elsayed et al. 1988; Izquierdo-Barrientos et al. 2013, 2015; Lin et al. 2021). However, a preliminary study of the size and density of the material must be conducted for fluidized leach reactors (Almendros-Ibáñez et al. 2019). The use of a fan is also necessary to ensure fluidization; this increases electricity consumption, and with it the cost to operate the system (Zondag et al. 2009). The use of fluidized bed reactors allows the heat transfer coefficients to be increased, as well as an increase in the sorption conversion rate compared to fixed bed reactors. However, few studies have been done regarding heat storage systems using the fluidized bed technology, due to its high pressure drop, the hydrodynamics of the complex reactor, as well as the modeling and erosion of the internal components (Zondag et al. 2009; Solé et al. 2015; Ströhle et al. 2016; Almendros-Ibáñez et al. 2019).

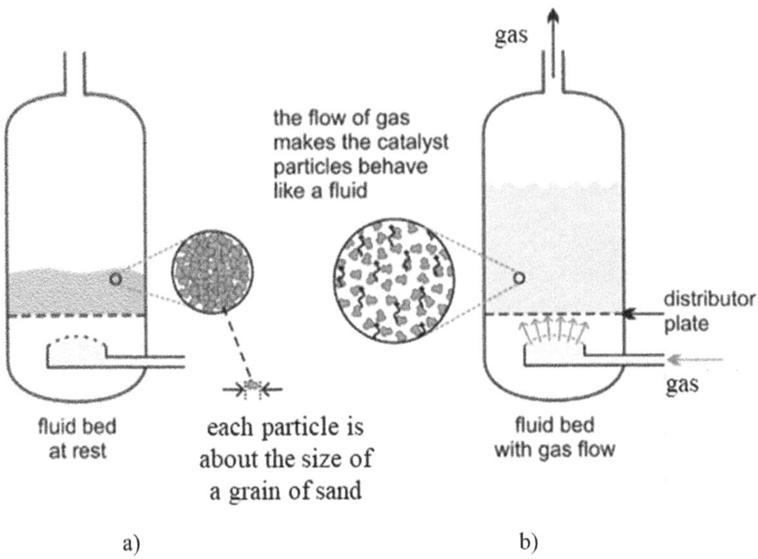

Figure 3.6. *Overview of the fluidized bed reactor: (a) at rest; (b) with a flow of gas (Lichtarowicz 2013)*

Zondag et al. (2009) have studied agitated reactors and gravity-assisted mass reactors (Figure 3.7). For both reactor systems, a large part of the electrical energy is consumed in transporting the solid. By optimizing the systems, high COP values can be obtained in both cases. However, the flow rate of the powder in the mixing reactor is considered to be more controllable with respect to the gravity-assisted mass reactor. The rotary mixing also allows for better penetration of the gas in the solid (Kunii et al. 1998).

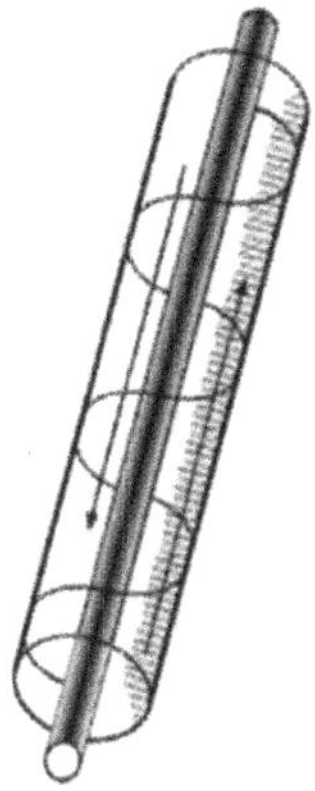

a)

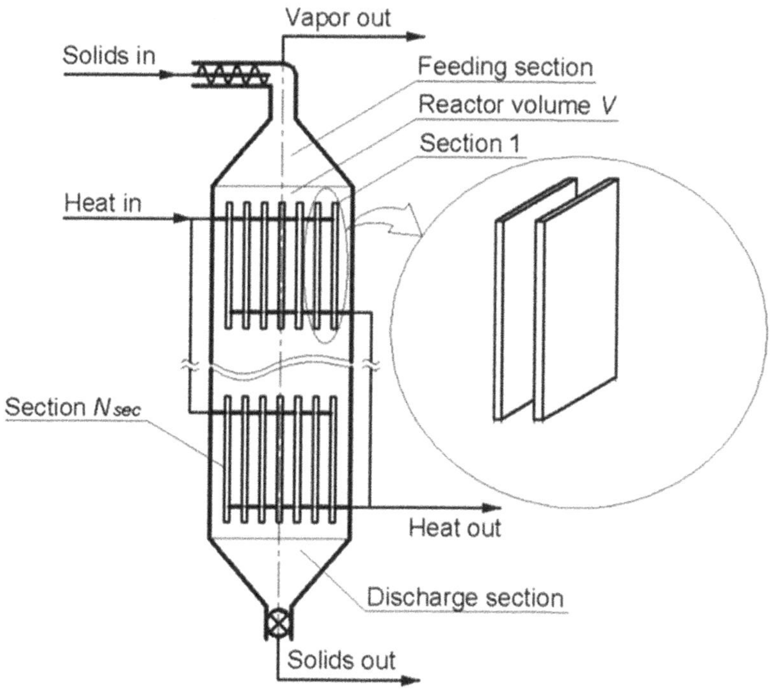

b)

Figure 3.7. *Overview of the reactors: (a) mixing; (b) gravity-assisted mass flow (Zondag et al. 2009)*

For these reasons, the mixing reactor is considered the most promising reactor. However, mechanical agitation can irreversibly damage the active solid material, which leads to increased system maintenance costs, high energy consumption and a more complex design. In addition, the controls of this type of system remains a challenge to be addressed (Scapino et al. 2017).

3.2.3. *Integration of storage systems in buildings*

3.2.3.1. *Strategy for charging the storage system*

The strategy for storage charging is strongly correlated with the source of heat.

Four possible solutions are quickly presented as follows:

– The heat used to charge the system can be produced by an industrial system (Guo et al. 2017). This generation provides little flexibility regarding the charging strategy. The *process* works according to very specific needs, and therefore adaptations must be made to fit these needs as much as possible. The system can then be recharged when the heat is produced by the *process* at a sufficiently high temperature.

– The heat can be produced by a solar thermal system. Once again, the flexibility remains low, because it is significantly constrained by solar lighting and the difficulty in predicting the amount of sunlight received. In the summer, it is possible to achieve long charging cycles at high temperatures. On the other hand, in winter, the cycles are much less frequent and are at much lower temperatures. The primary objective for control is therefore to optimize the solar proportions in order to produce heat at a sufficiently high temperature and for a sufficiently long time.

– A hybrid system can be envisaged; a system that couples both the auxiliary and an auxiliary source. The amount of energy produced by the auxiliary system could thus be partially recovered at the output of the storage system (Gondre 2016). The heat recovered can then be used to heat the building directly. Nevertheless, this approach is far from ideal from an exergetic point of view, since the auxiliary produces heat at high temperatures to then be used at low temperatures.

– All of the heat can be produced by a dedicated generator (electric heater, boiler, etc.). The objective in this case is to reduce the peak demand and the cost of heat production. Such a solution provides a good deal of flexibility at the network level and the control strategy can be easily adapted to the energy costs. However, it is necessary to recover the heat losses from the system for direct use in this case; otherwise, the ratio between the heat used during high-cost periods and the heat produced during low-cost periods will be too large to be made up for by the difference in energy costs.

3.2.3.2. *Strategy for discharging the storage system*

The discharge strategy can be considered based on different factors, such as the time or the load of the system, or depending on economic indicators, for example. Of course, this strongly depends on the intended application for the thermal storage system (Gondre 2016).

– If the goal is to optimize the proportion of solar energy used by the building, the heat will be used following the criterion of time, with more use at night when direct solar heating is not possible (Figure 3.8(a)).

– If the objective is to optimize the proportion of energy stored in the building, the storage system will be used to ensure a base level by using it as soon as the demand is present. In this case, it is not necessarily the goal to obtain significant coverage levels (Figure 3.8(b)).

– If the decision is to be made on economic considerations, heat is used as soon as the cost of energy exceeds a threshold value. It is generally assumed in the latter case that the cost of energy is highly correlated with the peak demand. However, it may differ from this period depending on local climatic conditions, internal loads, or the level of insulation of the building. In this case, the discharge strategy can be done in two different ways, which will depend on the strengths and weaknesses of the storage system:

- if the thermal power is the limiting factor, partially shedding the load allows for a smoothing of the energy demand, but the base production is always ensured by an auxiliary system (Figure 3.8(d));

- if the storage capacity (but not the power) is the limiting factor, fully shedding the load is more appropriate. When the demand for heat is above the threshold, all of these heating needs are then provided by the storage system, as shown in Figure 3.8(C).

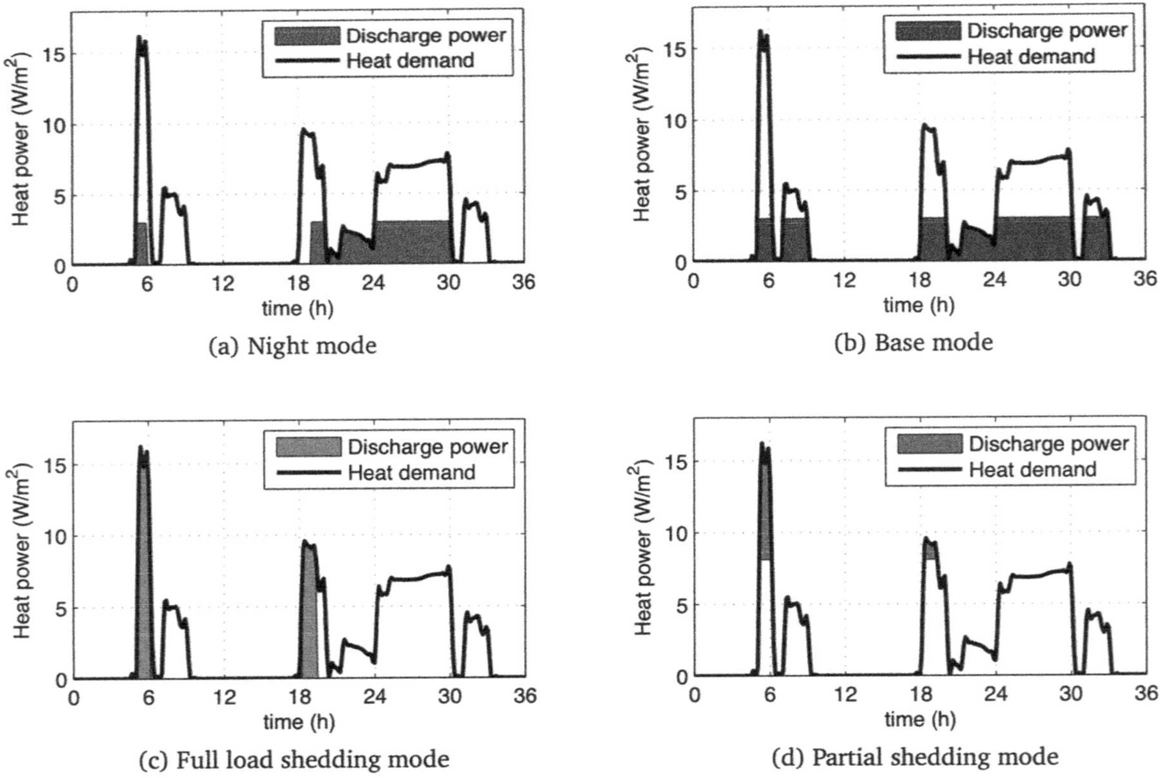

Figure 3.8. *Four possible strategies for controlling the storage discharge: (a) night mode; (b) basic mode; (c) full load-shedding mode; (d) partial load-shedding mode*

3.2.3.3. *Requirements in buildings*

In general, heat storage in buildings is mainly used for heating and/or the production of hot water for domestic use. The thermal and energy scales of such a system (peak power, amount of energy) can come up with significantly different values if the levels are determined only for heating, only for domestic hot water, or only or either one at the same time. Thus, the objective of section 3.2.3.3 is to provide some of the target values to be achieved when the system is scaled for use in buildings. More information about the data can be found in some previous studies (Bricka et al. 2011; Belz et al. 2015; Kuznik et al. 2015).

The heating demands of a building depend on a large number of factors. For this reason, within the framework of this chapter, it is difficult to give details on the calculation of the heating demands of buildings.

However, for more details, standards and guidelines have been established concerning the calculation of the heating demand for low and ultra-low consumption buildings. For example, using the sources mentioned above, it is possible to determine the values for the peak thermal power to be supplied by a heat storage system, as well as the amount of heat to be stored. Some examples are given in Table 3.1, involving buildings with very low energy consumption. A numerical model to evaluate these requirements has been created by Gondre et al. (2014).

Category	Maximum power $W.m^{-2}$	Annual requirements $kWh.m^{-2}.year^{-1}$
Low energy consumption building	16–24	20–30
Ultra-low energy consumption building	10–14	About 15

Table 3.1. *Maximum power and annual energy demand under the European standards and rules of the art according to Thullner (2010)*

The evaluation of the domestic hot water needs and the peak power required is more complicated to evaluate, since these values depend on the end user.

Schultz et al. (2008) have conducted a case study for a house with four residents, located in Sweden. The daily consumption of domestic hot water is 50 $L.D^{-1}$ per person. In this case, it has been shown that a 2 kW water heater was able to cover 82% of the needs, while a system of 3 kW made it possible to cover 86% of the demand. The annual energy consumption thus totaled 3 MWh for the four residents (Kuznik et al. 2015).

3.3. Existing prototypes of sorption heat storage

The existing experimental heat storage configurations are given in this section. The *focus* here is on laboratory prototypes, which therefore have a low Technology Readiness Level (TRL). Larger systems will be discussed in a later section.

3.3.1. *Closed systems*

3.3.1.1. *The MODESTORE Project*

The European MODESTORE project (Modular High Energy Density Heat Storage), a successor to the HYDES project, was piloted by AEE from the Austrian Institute of Sustainable Technologies (AEE INTEC) between 2003 and 2006 (Jaehnig et al. 2006). The sorption storage prototype was developed for the heating of a detached house.

The operation of this prototype is based on a closed system operating in a vacuum. The cylindrical reactor is filled with 200 kg of silica gel, consisting of an adsorber and a spiral heat exchanger (Figure 3.9).

The advantage of this type of design is that it is very compact, since all components are included within a single tank. However, despite the advantageous properties of silica gel and low pressure drops at the reactor, the experimental results have been far from the values that were expected. During the adsorption process, a temperature increase of about 5°C was observed, and the maximum power transferred was 400 W (Jaehnig et al. 2006). An energy density of 50 $kWh.m^{-3}$ (i.e. 50 $kWh.m^{-3}$) was reached, which is lower than the energy density of water for a temperature change of 70°C (~70 $kWh.m^{-3}$), or 25% of the total energy recoverable in this system (Bales et al. 2008).

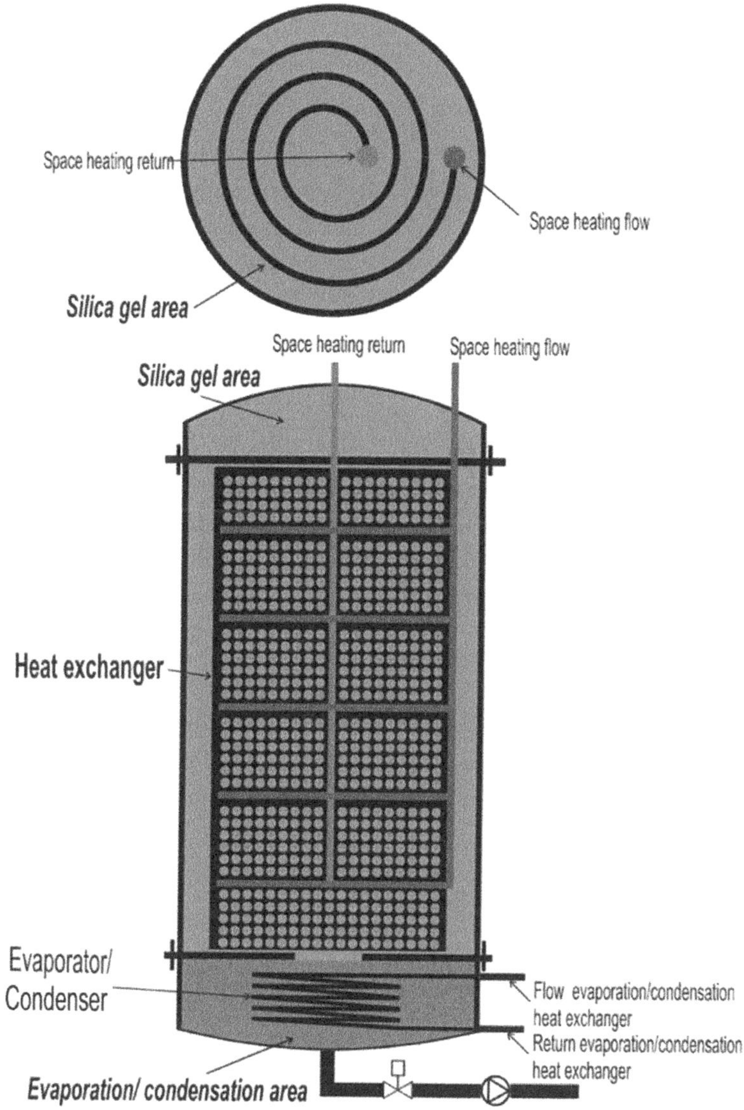

Figure 3.9. *MODESTORE heat storage prototype (Gartler et al. 2004). For a color version of this figure, see www.iste.co.uk/lepierres/heat2.zip*

These results demonstrate the low thermal conductivity of silica gel, as well as the low increase in temperature this material experiences during the adsorption process.

From this, it is concluded that the performances of the silica gel/water pairing are not adequate for heat storage.

3.3.1.2. *The E-HUB project*

The objective of the E-HUB project was to develop a 3 kWh sorption heat storage device for the heating of rooms at a temperature level of 40°C (de Boer et al. 2014; Finck et al. 2013, 2014). Zeolite 5A was chosen as the active material for this device because of its hygrothermal and mechanical stability, as well as for its non-corrosive and non-hazardous properties. Forty-one kilograms of material in the form of beads were placed in eight heat exchangers connected to the evaporator/condenser and operating at low pressure (Figure 3.10).

A maximum adsorption power of 971 W could be achieved for a desorption temperature of 103°C, condensation of 20°C, adsorption of 20°C and evaporation at 15°C. The maximum temperature rise was 31.2°C, with an average temperature rise of more than 20°C over a period of 1 h and 14 min. An average heating power of 800 W was obtained.

After 2 h, an amount of heat totaling to 1.4 kWh was transmitted into the circuit, and after 14 h, the total heat extracted from the reactor was 2.6 kWh, which represents 0.007 $Wh.g^{-1}$ of zeolite 5A. The temperature increase in the water circuit decreased to less than 0.5°C. The energy density of the material and of the system was calculated and represents 47 $kWh.m^{-3}$ and 12.5 $kWh.m^{-3}$, respectively.

3.3.1.3. *The COMTES project*

The European COMTES project was conducted by AEE at the Institute for Sustainable Technology, in Austria. This closed sorption storage system has been designed for providing heating and domestic hot water for a single-family house requiring 30 $kWh.m^{-3}$ (Köll et al. 2017). A diagram of the system is presented in Figure 3.11(a), equipped with 12 m^2 of vacuum tube solar collectors and two tanks containing 1 m^3 of zeolite 13X (Figure 3.11(b)). Different conditions were tested during the charging/discharging of the system. Under conditions deemed realistic though not optimal for the production of domestic heat, it has been determined to reach 150 $kWh.m^{-3}$. These conditions are a desorption temperature at 180°C and condensation at 17°C, as well as a sorption temperature at 45°C and evaporation at 20°C.

A maximum power of 4.65 kW was reached after 11 h of operation at 20°C, and the average power over the cycle was 2.8 kW.

a)

b)

Figure 3.10. *Adsorption/desorption module from the E-HUB project: (a) adsorption/desorption tank; (b) evaporator/condenser (Finck et al. 2014). For a color version of this figure, see www.iste.co.uk/lepierres/heat2.zip*

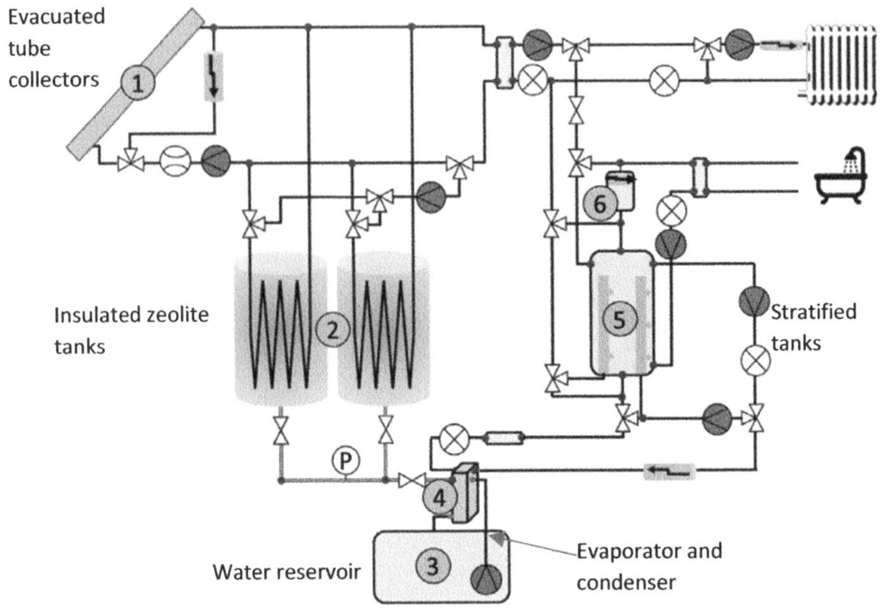

a)

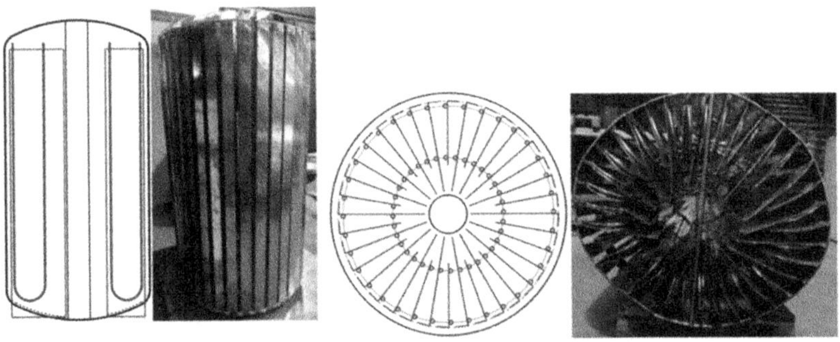

b)

Figure 3.11. *(a) The diagram of the closed system; (b) the heat exchanger containing the zeolite from the COMTES project (Köll et al. 2017). For a color version of this figure, see www.iste.co.uk/lepierres/heat2.zip*

3.3.1.4. *The CNR-ITAE compact heat storage prototype*

A prototype for sorption heat storage usable in mobile applications has been developed and tested at the CNR-ITAE in Italy (Palomba et al. 2017). Using 4.3 kg of AQSOA FAM Z02 zeolite, the prototype was designed and tested for cold storage, daily heat storage, as well as seasonal heat storage. The exchanger used for adsorption/desorption is in the form of flat tubes with fins whose exchange surface is 1.75 m^2 (Figure 3.12).

Different operating methods have been studied, as well as the effect of the operating conditions on the performance of the system. The authors concluded that the charging temperature has the most important influence on the system's performance. For heat storage for everyday use, the prototype developed is capable of storing 620 Wh at 75°C for desorption and 1,100 Wh at 9°C, which represents an average energy density of 280 $Wh.kg^{-1}$ and an average power during the discharge phase ranging from 200 to 650 W. The results show that the performance and efficiency are comparable to those found by other authors (Girnik and Aristov 2016; Van Alebeek et al. 2018). This prototype is noteworthy in that it allows low-temperature waste heat (<100°C) to be used for both heat and cold storage.

3.3.2. *Open systems*

3.3.2.1. *The SPF project*

At the Fuer Solartechnik Institute (FPS) in Switzerland, a test was conducted on a prototype containing 7 kg of 13X zeolite (Bales et al. 2008). It demonstrated a good storage capacity of 1 kWh for an output power of 800 W at a temperature of 55°C. The storage capacity is a nonlinear function of the load temperature. Note that 57% of the storage capacity at 180°C is reached at 110°C, while 94% is reached by 150°C. This is a good point for solar applications, because a high charging temperature is not always available.

The maximum temperature of 70°C was reached for a sorption temperature of 20°C and a water vapor pressure of 2.34 kPa.

a)

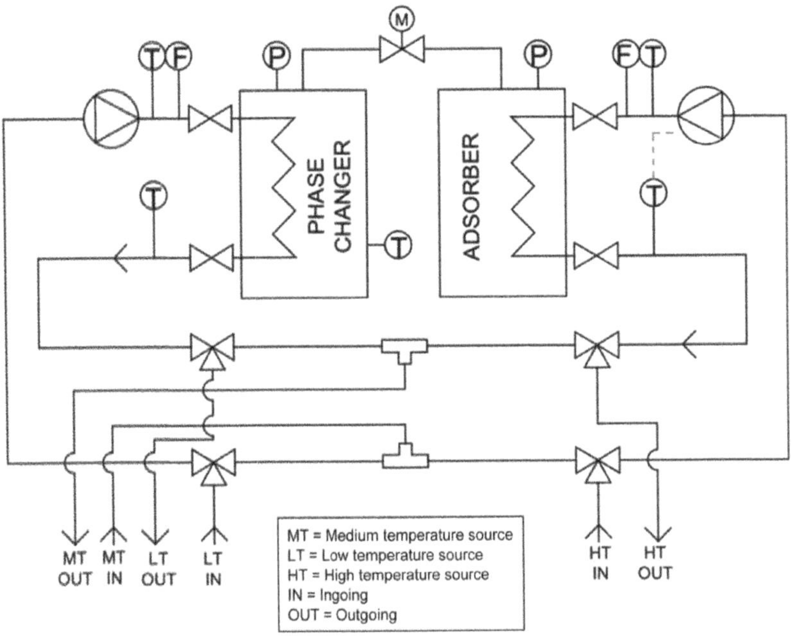

b)

Figure 3.12. *Schematic overview of the heat storage system developed by CNR-ITAE (Palomba et al. 2017). For a color version of this figure, see www.iste.co.uk/lepierres/heat2.zip*

3.3.2.2. *The MonoSorp project*

In the MonoSorp project (developed by the Institute of Thermodynamics and Thermal Engineering (ITW) at the University of Stuttgart), a 12 kWh prototype was developed for heating (Bales et al. 2008). A 7.85 m^3 system was scaled to meet the heating needs of a single family house. For this prototype, a mass of 70 kg of zeolite 4A was used. The material has a good storage density of 160 $kWh.m^{-3}$ and a very good compatibility between the volume of the material and the volume of the prototype.

a)

b)

Figure 3.13. *MonoSorp Prototype (Bales et al. 2008). For a color version of this figure, see www.iste.co.uk/lepierres/heat2.zip*

To improve the adsorption kinetics and to reduce thermal losses, a honeycomb structure divided into monoliths was designed instead of an ordinary filling (Figure 3.13).

During the charging, the air entering the zeolite reactor is preheated up to 180°C by the solar collector.

During the discharge, internal air at about 19°C and a water vapor pressure of 0.87 kPa is sent through the zeolite stock until it is saturated. A temperature rise of 22°C was reached. The reactor provided a maximum power of 1,000 W, but despite these encouraging results, the discharge power remains quite low, on the order of 1–1.5 kW. In addition, the necessary regeneration temperature of 180°C is quite difficult to reach with standard solar powered systems.

3.3.2.3. *The Flow-TCS project*

A prototype for a rotary drum reactor was developed at the Austrian Solar Innovation Center (ASIC), with the goal of improving heat and mass transfers through the continuous rotation of the sorption material (Zettl et al. 2014). The diameter of this reactor is 0.7 m with a depth of 0.3 m and the capacity to contain 50–53 kg of type 4A and 13X zeolite in the form of beads (Figure 3.14). Desorption temperatures of 180°C and 230°C were applied for zeolite 4A and 230°C for zeolite 13X. During the adsorption, the air temperature at the inlet was maintained at 25°C, and the water vapor content remained constant on each adsorption process but from 9 to 19 $g_{H2O}.kg_{air}^{-1}$.

The authors concluded that there were several positive points about the rotary drum reactor compared to other reactors: its high specific thermal power, process control, low pressure drops at the sorption bed and its high coefficient of performance (COP). The energy densities of zeolite 4A and 13X have been calculated at 153 and 147 $kWh.m^{-3}$ respectively, with a maximum temperature rise of 36°C obtained for the 13X zeolite dehydrated at 230°C. The maximum output power is 30 $W.kg^{-1}$, with an average power of 13–20 $W.kg^{-1}$.

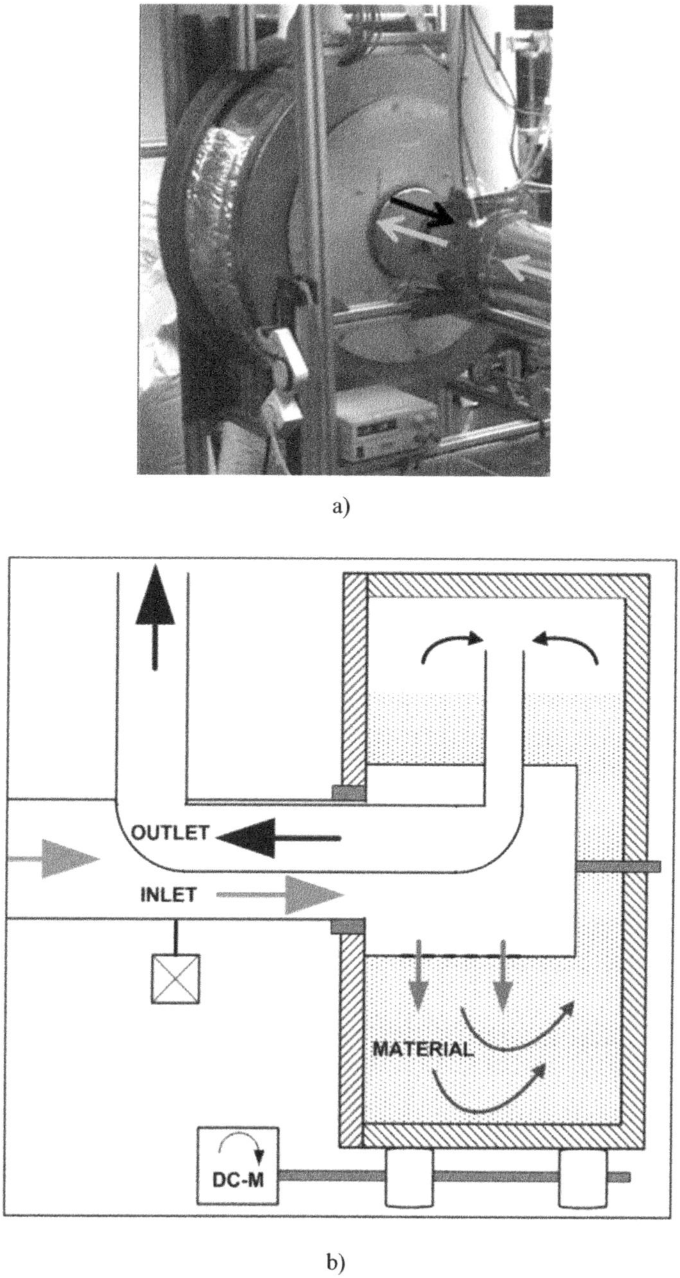

Figure 3.14. *Prototype developed for the project Flow-TCS (Zettl et al. 2014). For a color version of this figure, see www.iste.co.uk/lepierres/heat2.zip*

3.3.2.4. *The ZAE Bayern project*

ZAE Bayern developed a mobile prototype of an open heat storage system (Krönauer et al. 2015). To carry out this study, two reactors containing 14 tons of 13X zeolite were built. The principle of this mobile system consists of a charging phase carried out on a waste incineration station equipped with a steam turbine, and then once charged, the reactor is transported to the discharge station (Figure 3.15).

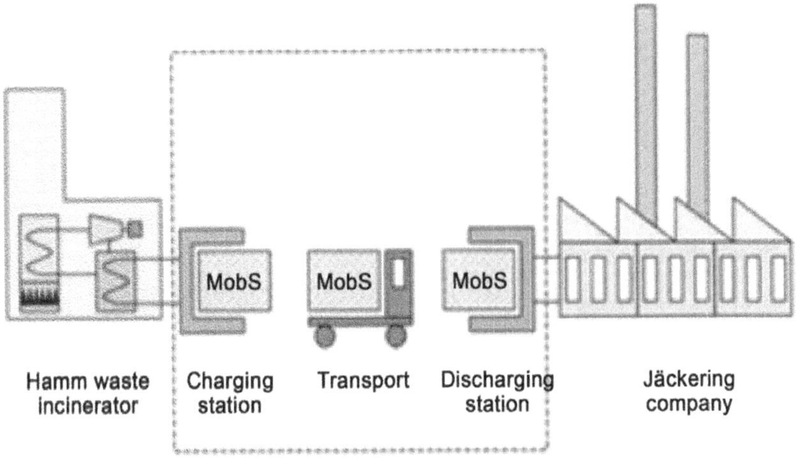

a)

b)

Figure 3.15. *Mobile heat storage prototype designed by ZAE Bayern (Krönauer et al. 2015)*

To charge the system, ambient air heated up to 130°C is transferred into the reactor. After leaving the reactor, the air temperature is between 60 and 120°C, which allows for this heat to be reused. During the discharge phase, moist air preheated to 60°C is introduced into the reactor. Upon leaving the reactor, the temperature measured reached 160°C and the energy density of the system was 103 $kWh.m^{-3}$, thus the density of the material was 630 $kg.m^{-3}$.

This reproduction power value is very low compared to other existing systems (Bales et al. 2008; Zettl et al. 2014; Van Alebeek et al. 2018; Finck et al. 2013), which is explained by the poor distribution of the gas flow through the zeolite bed, the variation in humidity during the cycles and also by the fact that the discharge system within the system does not work in stationary conditions. Thus, the discharge phase was not carried out all the time at maximum power.

However, the authors found that this system model prevents the release of 616 kg of CO_2 into the atmosphere per cycle. After 99 cycles, despite the mechanical stress due to transport, no degradation of the material was observed. The energy cost has been calculated at around €73 MWh^{-1}, and the authors claim that by optimizing the maintenance of the storage system as well as increasing the quantity of the material on the order of 50 tons, this price value can be reduced by up to €30 MWh^{-1}, making it competitive with the primary energy costs of oil or gas.

3.3.2.5. *The Smart Energy Regions Brabant project program*

At Eindhoven University of Technology laboratory in the Netherlands, a modular prototype has been developed, containing four modules with a total capacity of 250 liters (about 170 kg) (Figure 3.16(a)) (Gaeini et al. 2017, 2018; Van Alebeek et al. 2018). Each module of the reactor contains 62.5 liters of 13X zeolite (Figure 3.16(b)).

This storage system was designed and built for heating and providing domestic hot water for buildings. The modular reactor has been designed with the goal of reducing pressure losses by reducing the height/width ratio of the bed in order to maximize the efficiency of the reactor. A fixed bed reactor was chosen because of its simplicity and low cost.

a)

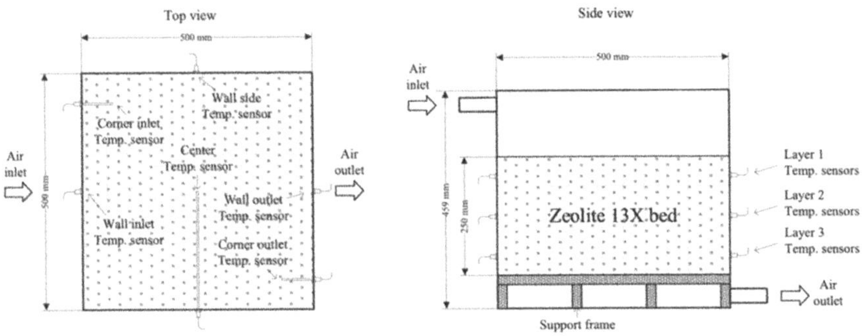

b)

Figure 3.16. *Prototype of the Smart Energy Regions Brabant program: (a) complete system; (b) modular reactor (Gaeini et al. 2018; Van Alebeek et al. 2018). For a color version of this figure, see www.iste.co.uk/lepierres/heat2.zip*

For heating a building, during the system charge, a temperature of 190°C was used, and for the discharge of the system, a water vapor pressure of about 1 kP was maintained at the temperature of 10°C. The energy density has been calculated for the material and for the reactor; these values are 198 kWh.m^{-3} and 108 kWh.m^{-3}, respectively. Each module can be charged with 17 kWh of energy for 7 h. The total energy obtained during the discharge phase is approximately 12 kWh, with a maximum power per module of 0.9 kW, which represents a total power of 3.6 kW over 10 h, as compared with the target value of 4 kW. During these 10 h, 70–75% of the total energy is released with a constant high power, and the rest is released during the final 5–8 h of the discharge process, during which the power gradually decreases.

For the production of domestic hot water, a desorption temperature of 180°C was used, while the adsorption was carried out at 13°C at a water vapor pressure of about 1.3 kPa. The results showed that it is possible to store 54 kWh of thermal energy, providing a power of 4.4 kW. The efficiency of this system is 76% and can be increased up to 91% if the sensitive part of the reactor material is recovered. The energy densities for the material and for the reactor were calculated at 112 kWh.m^{-3} and 61 kWh.m^{-3} respectively.

The authors claimed that this system is a good compromise for providing both heat and domestic hot water to the building. Certain points can be improved in order to increase the system's efficiency. For example, the sensitive heat of the solid material in the reactor can be recovered. The greatest pressure drop in the above configuration is located at the level of the heater, which must be replaced by an air-to-air heat exchanger in order to minimize this pressure drop. The discharged air at the inlet of the reactor can be pre-heated to increase the temperature rise in the reactor.

3.3.2.6. *The STAID project*

For the STAID Project (for inter-seasonal storage of thermal energy in buildings), supported by CETHIL at INSA Lyon (France), another open-type sorption heat storage system based on 13X zeolite was developed (Johannes et al. 2015; Tatsidjodoung et al. 2016). The goal of this study was to develop a prototype to supply at least 2 kW over 2 h for domestic use during peak consumption. The prototype consists of two separate reactors containing 40 kg of zeolite each, which can be connected to each other in parallel or in series (Figure 3.17).

A series of tests was carried out for different operating conditions during charging and discharging: charging at 120°C or 180°C and discharging at 20°C and different water vapor pressures. The charging results show that a storage density of 104 kWh.m^{-3} is obtained with a maximum output power of 2.25 kW (27.5 W.kg^{-1}) for 6 h. This value is greater than the required 2 h of operating time. A temperature increase of 57°C was achieved for an energy density of the material of 115 kWh.m^{-3}. The authors found that the configuration of the reactors in series provided a more constant thermal power than a configuration in parallel.

Figure 3.17. *The prototype of the STAID project (Johannes et al. 2015). For a color version of this figure, see www.iste.co.uk/lepierres/heat2.zip*

3.3.2.7. *The SolSpaces project*

As part of the SolSpaces project in Germany (Weber et al. 2016), a new solar heating system was developed with the purpose of providing all energy needs for low-consumption buildings. In this project, the seasonal sorption thermal energy storage system is a central element.

The volume of the storage reactor was calculated according to the heating needs of each of the experimental buildings in SolSpaces. The building was equipped with a conventional heating element. A sorption tank of 4.3 m^3 was needed, providing 700 kWh of energy. This tank was filled with beads of 13X zeolite. The total volume of the storage system, including insulation, is 8 m^3. The reservoir is divided into four quadrants, and each quadrant is subdivided into six segments (Figure 3.18). The flow of air enters the tank through the central duct, passes through each segment, and exits through one of the four ducts on the vertical edges. Two-segment groupings are stacked on top of each other to form pairs, and each pair of segments can operate individually, allowing each two segment modules to be controlled separately for the charging and discharging phases. In order to obtain detailed data of the overall system, the temperature, humidity, solar radiation and air flow, along with the electrical energy used by the system for devices such as fans, were recorded in several places in the storage tank, in the entire system and in the building.

a)

b)

Figure 3.18. *The sorption thermal energy storage tank in the SolSpaces project (Weber et al. 2016). For a color version of this figure, see www.iste.co.uk/lepierres/heat2.zip*

The operating conditions of the storage system are as follows: a water vapor pressure of 1 kPa kept constant during the charging and discharging phase, while the temperature of 180°C was used during the charging phase and 20°C during the discharging phase. A recording of data over a 13-h period showed small drops in pressure in the reactor, while the thermal adsorption power turned out to be in the range of 790–565 W, and the difference in humidity between the inlet and the outlet of the sorption tank decreased by 8.8–6.3 $g.kg^{-1}$. The average temperature obtained was on the order of 52°C, a good value for a single family house.

The authors justified the low thermal power levels by the fact that the humidity of an inhabited building is higher than that of the uninhabited building that was used in the SolSpaces project. The authors also mentioned that the thermal power can be increased by adapting the air flow.

3.4. System performances: an analysis of the prototypes presented

A summary of the low temperature long-term sorption heat storage systems is given in Table 3.2. An analysis is made of the energy density, the maximum power, the sorption and desorption temperatures, as well as the rise in maximum temperatures. Comparing the systems presented above is obviously difficult, given that there is no single approach to designing and testing these systems. In addition, the systems are often tested under ideal conditions for sorption materials rather than under realistic application conditions. For example, the desorption temperature of 180–230°C is rarely achievable for discharge within a building under real-world conditions.

On the basis of the literature presented, it can be observed that the systems are at differing levels of development in terms of their scale of operation and technological maturity. Therefore, it is difficult to compare sorption heat storage systems based solely on the energy storage density or on the discharge power of the project/institution system.

In terms of materials, zeolite is the most frequently studied material for long-term storage. As described in Chapter 1, this material is favored due to the fact that it has the following properties:

– good mechanical strength;

– a large volume of pores;

– a large surface area;

– high adsorption capacity;

– good transporting of heat and mass.

Project/ institution	Year	Active material	Quantity, kg	Type of system	E, $kWh.m^{-3}$	P Discharge, kW	T desorp, °C	T sorp, °C	ΔT maxsorp, °C	PH2O, kPa	T discharge, h	References
MODESTORE	2006	Silica gel	200	Closed	50^{m}	0.5–1	88	–	70	1.9–3.5	–	(Jaehnig et al. 2006; Bales et al. 2008; Scapino et al. 2017)
E-HUB	2014	Zeolite 5A	41	Closed	47^{m} 12^{s}	0.8	103	20	51	23	14	(de Boer et al. 2014; Finck et al. 2014)
COMTES	2017	Zeolite 13X	$\sim680^{A}$	Closed	178^{m}	2.1	180	45	25	0.87–3.35	11	(Köll et al. 2017)
CNR-ITAE	2017	Zeolite AQSOA Z02	4.3	Closed	280 $Wh.kg^{-1}$	0.62–1.1	75–90	30–40	55	0.9–1.7	0.83	(Palomba et al. 2017)
SPF	2006	Zeolite 13X	7	Open	180^{m} 58^{s}	0.8–1.8	180	–	70	2.3	–	(Bales et al. 2008; Scapino et al. 2017)
MonoSorp	2006	Zeolite 4A	70	Open	160^{m} 120^{s}	1–1.5	180	20	42	0.9	–	(Bales et al. 2008)

Table 3.2. *Comparison of low-temperature long-term heat storage systems. Exponents: A = calculated by the authors; m = material; s = system*

Project/ institution	Year	Active material	Quantity, kg	Type of system	E, $kWh.m^{-3}$	P Discharge, kW	T desorp, °C	T sorp, °C	ΔT maxsorp, °C	PH2O, kPa	T discharge, h	References
Flow-TCS	2014	Zeolite 4A Zeolite 13X	53 50	Open Open	153^s 95^s	1.5 1.5	230 180	25 25	84 44	2.5 1.2	~10 ~10	(Zettl et al. 2014)
ZAE Bayer	2015	Zeolite 13X	14,000	Open	103^m	125	130	60	160	19.9	12	(Krönauer et al. 2015; Scapino et al. 2017)
SMART Energy	2017	Zeolite 13X	170	Open	198^m 108^s	17	190	10	45	1	7	(Gaeini et al. 2017; Van Alebeek et al. 2018)
STAID	2015	Zeolite 13X	80	Open	115^m	2.25	120–180	20	57	1.6	6–14	(Johannes et al. 2015; Tatsidjodoung et al. 2016)
SolSpaces	2014	Zeolite 13X	$2,900^A$	Open	–	0.79–0.57	180	20	42	1	13	(Weber et al. 2016)

Table 3.2. *Comparison of low-temperature long-term heat storage systems. Exponents: A = calculated by the authors; m = material; s = system*

The other indicators to be considered in order to reach a conclusion on the performance and suitability of the system for use in heat storage application are a stable thermal power over time, the thermal losses of the system during the charging/discharging phases as well as during the cycles, the stability of the materials chosen for the system, the size, the geometry of a reactor, the sorption temperature and the water vapor pressure during the discharge phase. For example, the simplest geometrical forms are often the best, since they allow for the volume of material to be maximized relative to the volume of the prototype. The MonoSorp project is a good example of a highly compact design. However, additional developments are necessary to optimize the heat exchange inside the reactor (Bales et al. 2008). From among the systems for heat storage by adsorption studied and presented, the ZAE Bayern project is at a relatively advanced stage of development. The system has two large reactors with an energy density of 103 $kWh.m^{-3}$. This system is run under real operating conditions. The results obtained are very promising, yet nevertheless, this system is not suitable for home heating or for domestic hot water. The system requires high desorption temperatures, and the adsorption temperature of 160°C was achieved under specific conditions: air preheated at 60°C with a water vapor pressure of 20 kPa. These factors make it difficult to incorporate this system into buildings, but it is a good example of a system that can be integrated into an industrial process.

Rotating prototypes or complex heat exchangers do not necessarily bring the expected improvement to the performance of the system. Closed sorption systems are very promising in small-scale experiments, but these results have not been confirmed on larger scale prototypes. Modular reactors allow for the pressure drops in the system to be reduced, and therefore provide improved performances. Separate reactors that can efficiently transport materials within the system can increase its performance, but there still remain technological bottlenecks in the application of adsorption heat storage.

3.5. The influence of kinetics

Heat storage by adsorption is an interesting solution for short- and long-term thermal energy storage. However, the technological maturity of these systems remains very low compared to sensitive and latent heat storage systems. The main obstacle is the kinetics of the transfer of heat and mass.

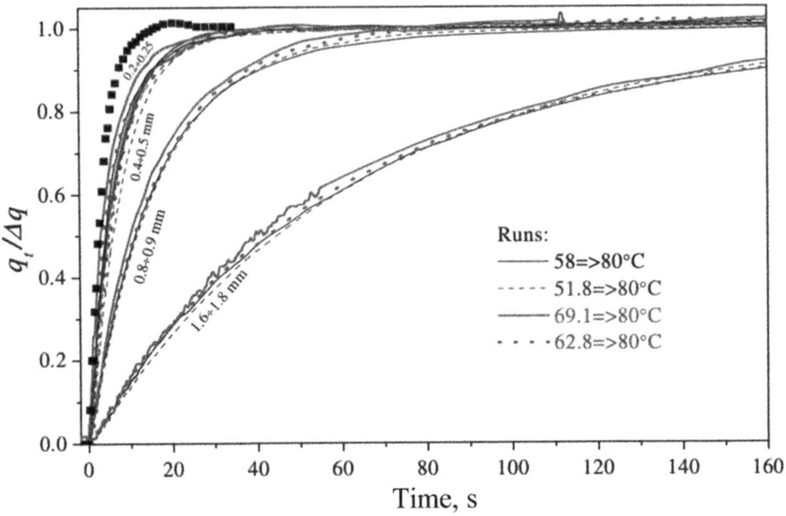

a)

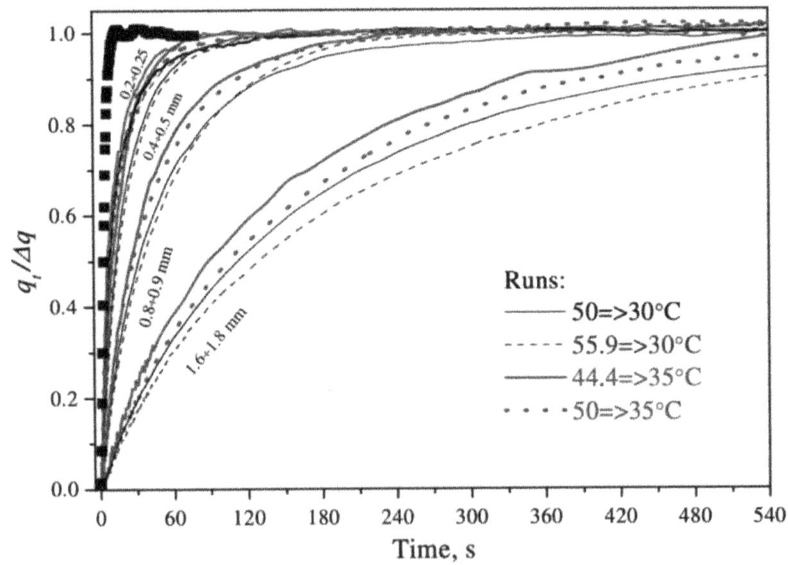

b)

Figure 3.19. *Curves showing (a) the desorption and (b) the adsorption for different grain sizes (Glaznev and Aristov 2010). For a color version of this figure, see www.iste.co.uk/lepierres/heat2.zip*

In order to achieve optimal adsorption kinetics, several parameters must be taken into account to improve the design of the heat and mass exchanger: the adsorption kinetics, the contact surface of the adsorbent with the adsorbate, the conductivity, the permeability, the exposure time, the flow rate of the adsorbent and the dimensions of the reactor.

The influence of the sorbent particle size and the thickness of the sorbent layers and the performance of closed and open sorption systems have been studied by N'Tsoukpoe et al. (2014). The authors were able to conclude that the grain size is a powerful tool for controlling the adsorption/desorption dynamics. For example, a study conducted on the different ranges of grain sizes of Fuji silica RD shows that the speed is highly dependent on the grain size (Figure 3.19) (Glaznev and Aristov 2010).

It can be concluded that the specific power is more dependent on the particle size of the solid material than on the conditions applied. The power of an open system drops when it contains powder or small particles, which may require a higher air flow in order to provide satisfactory power. However, the increase in the pressure drop necessarily leads to additional electrical consumption for the fans or compressors, which increases the operating cost.

The granular medium used has an influence on the heat and mass transfer. The low heat transfer in the bed of the material, especially for closed systems operating in a vacuum, is due to the low conductivity of the solid adsorbent, as well as to the low contact levels between the grains and the walls of the reactor (N'Tsoukpoe et al. 2014). The mass transfer in the bed depends on the permeability and the thickness of the bed. High permeability requires high porosity; though high porosity does not necessarily mean there is good permeability (N'Tsoukpoe et al. 2014). Thus, it is necessary to measure the permeability.

For the same energy density, it is possible to increase the permeability thanks to an optimal grain size or particle distribution (Michel et al. 2012). The choice of the shape of the grains plays an important role in the configuration of a storage system. For example, the grains have a better adsorbate diffusion, but the grains only have a light contact with the walls of the tank, which decreases the heat transfer.

3.6. Real-scale systems

This section focuses on existing experimental systems whose technological maturity is greater than 7. It is interesting to note that very few systems fall into this category of systems operating under real-world conditions. This shows the clear potential for development of this technology.

In Munich, Germany, a large–scale experiment (at 1,350 kWh) under real operating conditions has been ongoing since 1997. This system was designed to meet the energy needs of a 1,625 m^2 school during its peak consumption period of a few hours with a volume flow of 6,000 m^3.h^{-1}. This installation also has the goal of providing cooling for a jazz club at the same time, which has a reasonable cooling load. To accomplish this, the system that was installed includes energy storage of 7 tons of 13X zeolite. The charging of this storage is carried out at night using the urban heat network, a heat network which provides steam at 140°C. This allows the system to store up to 124 kWh.m^{-3} of heat. The nominal power offered by the system is 130 kW (18.6 W.kg^{-1}) (Hauer 2007b). A photo of the installation is shown in Figure 3.20.

Figure 3.20. *Storage system installed in Munich, Germany (Hauer 2007b)*

In 1998, the European Union financed the High Energy Density Sorption Heat Storage for Solar Space Heating (HYDES) project (Nunez et al. 2003; Hauer 2007b). The objective of this project was to develop an interseasonal heat storage system based on a closed cycle using silica gel. With this in mind, a pilot was built by the company AEE INTEC (see Figure 3.21); its 20.4 m^2 solar thermal collectors allowed it to recharge the storage in sunny weather. The system was designed for an apartment with floor and wall heating at 40/30°C, and with a maximum heating power of 2.2 kW. The experimental results show that the energy density of the developed system is 20% lower than the scaled value of 150 kWh.m^{-3} silica gel.

Figure 3.21. *Installation of the HYDES project in Gleisdorf, Austria (Hauer 2007b). For a color version of this figure, see www.iste.co.uk/lepierres/heat2.zip*

3.7. Conclusion

The use of sorption in thermal energy storage, based on solid/water vapor adsorption, is a promising approach for the efficient use of renewable energy sources, such as solar energy, or residual heat from the industrial, domestic and transport sectors. This chapter presents a discussion of adsorption thermal energy storage systems.

First of all, the principles of operation of adsorption materials are described, followed by a discussion of the configuration of a solid/gas adsorption system and the principle of operation of an adsorption cycle. A classification is given for systems and reactors, including closed or open systems, reactors integrated or separated from the storage tank, as well as of the types of particle beds in the reactor and the operation of the reactors. Open adsorption systems offer certain advantages, in particular their ability to operate at atmospheric pressure and ambient temperature, their greater heat and mass transfer levels, their simple configuration and the fact that they are easy to manufacture and maintain. While iron adsorption systems can use various different work fluids, they are better suited to different climates and can combine the functions of cooling in summer with heating in winter. It has been shown that the performance of the storage material, and therefore the ability to take advantage of the potential it offers, depend on the design of the process, in which the configuration of the separate reactors with the transporting of materials is more efficient and can increase the performance of the system. However, this type of system remains a challenge to overcome. Modular reactors, particularly in open systems, should be given preference so as to limit pressure losses. A description is given of the integration of storage systems into buildings and the requirements for such systems. Finally, the main prototypes and their performances, as well as the existing experimental systems, are examined.

A demonstration of heat storage systems by adsorption has shown the great potential of the technology. However, the development of the system has yet to go beyond laboratory applications. Difficulties must still be overcome and problems remain to be solved in order for this technology to be usable at an industrial level.

More in-depth research on the operating reactors is needed. The most noteworthy efforts in this sense are focused on the development of new adsorption materials, increasing the heat storage density under the operating conditions of the systems, improving heat and mass transfers at the level of components and systems.

3.8. References

Abedin, A.H. (2011). A critical review of thermochemical energy storage systems. *TOREJ*, 4, 42–46. doi: 10.2174/1876387101004010042.

Abedin, A.H. and Rosen, M.A. (2012). Closed and open thermochemical energy storage: Energy- and exergy-based comparisons. In *Energy: 23rd International Conference on Efficiency, Cost, Optimization, Simulation and Environmental Impact of Energy Systems, ECOS 2010*, 41. doi: 10.1016/j.energy.2011.06.034.

van Alebeek, R., Scapino, L., Beving, M.A.J.M., Gaeini, M., Rindt, C.C.M., Zondag, H.A. (2018). Investigation of a household-scale open sorption energy storage system based on the zeolite 13X/water reacting pair. *Applied Thermal Engineering*, 139, 325–333. doi: 10.1016/j.applthermaleng.2018.04.092.

Almendros-Ibáñez, J.A., Fernández-Torrijos, M., Díaz-Heras, M., Belmonte, J.F., Sobrino, C. (2019). A review of solar thermal energy storage in beds of particles: Packed and fluidized beds. *Solar Energy*, 192, 193–237. doi: 10.1016/j.solener.2018.05.047.

Bales, C., Gantenbein, P., Jaenig, D., Kerskes, H., Summer, K., Essen, M., Weber, R. (2008). Laboratory tests of chemical reactions and prototype sorption storage units. Report B4 of Subtask B, IEA Solar Heating and Cooling Programme [Online]. Available at: https://www.iea-shc.org/data/sites/1/publications/task32-b4.pdf [Accessed 24 March 2021].

Belz, K., Kuznik, F., Werner, K.F., Schmidt, T., Ruck, W.K.L. (2015). Thermal energy storage systems for heating and hot water in residential buildings. In *Advances in Thermal Energy Storage Systems*, Cabeza, L.F. (ed.). Woodhead Publishing, Sawston. doi: 10.1533/9781782420965.4.441.

Bennici, S., Polimann, T., Ondarts, M., Gonze, E., Vaulot, C., Le Pierrès, N. (2020). Long-term impact of air pollutants on thermochemical heat storage materials. *Renewable and Sustainable Energy Reviews*, 117, 109473. doi: 10.1016/j.rser.2019.109473.

de Boer, R., Vanhoudt, D., Claessens, B., De Ridder, F., Reynders, G., Cuypers, R., Oversloot, H., Finck, C., van Soest, C., de Jong, A.J., et al. (2014). Energy-hub for residential and commercial districts and transport – D3.2. Report on a combination of thermal storage techniques and components.

Bricka, V., Kuznik, F., Johannes, K., Virgone, J. (2011). Evaluation of thermal energy storage potential in low-energy buildings in France. In *Proceedings of the ISES Solar World Congress 2011. Presented at the ISES Solar World Congress 2011, International Solar Energy Society*. doi: 10.18086/swc.2011.15.03.

Elsayed, M.M., Megahed, I.E., El-Refaee, M.M. (1988). Experimental testing of fluidized bed thermal storage. *Solar & Wind Technology*, 5, 15–25. doi: 10.1016/0741-983X(88)90086-0.

Elsheniti, M.B., Elsamni, O.A., Al-dadah, R.K., Mahmoud, S., Elsayed, E., Saleh, K. (2018). Adsorption refrigeration technologies. In *Sustainable Air Conditioning Systems*, Ghenai, C. and Salameh, T. (eds). IntechOpen, London, 71–95. doi: 10.5772/intechopen.73167.

Finck, C., Spijker, H., Jong, A., Henquet, E., Oversloot, H., Cuypers, R. (2013). Design of a modular 3 kWh thermochemical heat storage system for space heating application. *2nd International Conference on Sustainable Energy Storage*, 19–21 June, Dublin.

Finck, C., Henquet, E., van Soest, C., Oversloot, H., de Jong, A.J., Cuypers, R., Spijker, H. (2014). Experimental results of a 3 kWh thermochemical heat storage module for space heating application. *Energy Procedia*, 48, 320–326. doi: 10.1016/j.egypro.2014.02.037.

Fumey, B., Weber, R., Baldini, L. (2019). Sorption based long-term thermal energy storage – Process classification and analysis of performance limitations: A review. *Renewable and Sustainable Energy Reviews*, 111, 57–74. doi: 10.1016/j.rser.2019.05.006.

Gaeini, M., Wind, R., Donkers, P.A.J., Zondag, H.A., Rindt, C.C.M. (2017). Development of a validated 2D model for flow, moisture and heat transport in a packed bed reactor using MRI experiment and a lab-scale reactor setup. *International Journal of Heat and Mass Transfer*, 113, 1116–1129. doi: 10.1016/j.ijheatmasstransfer.2017.06.034.

Gaeini, M., van Alebeek, R., Scapino, L., Zondag, H.A., Rindt, C.C.M. (2018). Hot tap water production by a 4 kW sorption segmented reactor in household scale for sea-sonal heat storage. *Journal of Energy Storage*, 17, 118–128. doi: 10.1016/j.est.2018.02.014.

Gartler, G., Jähnig, D., Purkarthofer, G., Wagner, W. (2004). Development of a high energy density sorption storage system. *Proceedings of Eurosun 2004*.

Girnik, I.S. and Aristov, Y.I. (2016). Dynamic optimization of adsorptive chillers: The "AQSOA™-FAM-Z02 – water" working pair. *Energy*, 106, 13–22. doi: 10.1016/j.energy.2016.03.036.

Glaznev, I.S. and Aristov, Y.I. (2010). The effect of cycle boundary conditions and adsorbent grain size on the water sorption dynamics in adsorption chillers. *International Journal of Heat and Mass Transfer*, 53, 1893–1898. doi: 10.1016/j.ijheatmasstransfer.2009.12.069.

Gondre, D. (2016). Numerical modeling and analysis of heat and mass transfers in an adsorption heat storage tank: Influences of material properties, operating conditions and system design on storage performances. PhD Thesis, Université de Lyon, Lyon.

Gondre, D., Johannes, K., Kuznik, F. (2014). Specification requirements for inter-seasonal heat storage systems in a low energy residential house. *Energy Conversion and Management*, 77, 628–636. doi: 10.1016/j.enconman.2013.09.061.

Guilleminot, J.J., Choisier, A., Chalfen, J.B., Nicolas, S., Reymoney, J.L. (1993). Heat transfer intensification in fixed bed adsorbers. *Heat Recovery Systems and CHP*, 13, 297–300. doi: 10.1016/0890-4332(93) 90052-W.

Guo, S., Zhao, J., Wang, W., Yan, J., Jin, G., Wang, X. (2017). Techno-economic assessment of mobilized thermal energy storage for distributed users: A case study in China. *Applied Energy*, 194, 481–486. doi: 10.1016/j.apenergy.2016.08.137.

Hauer, A. (2007). Sorption theory for thermal energy storage. In *Thermal Energy Storage for Sustainable Energy Consumption*, Paksoy, H.Ö. (ed.). Springer, Dordrecht, 393–408, 409–427. doi: 10.1007/978-1-4020-5290-3_24.

Izquierdo-Barrientos, M.A., Sobrino, C., Almendros-Ibáñez, J.A. (2013). Thermal energy storage in a fluidized bed of PCM. *Chemical Engineering Journal*, 230, 573–583. doi: 10.1016/j.cej.2013.06.112.

Izquierdo-Barrientos, M.A., Sobrino, C., Almendros-Ibáñez, J.A. (2015). Energy storage with PCM in fluidized beds: Modeling and experiments. *Chemical Engineering Journal*, 264, 497–505. doi: 10.1016/j.cej.2014.11. 107.

Jaehnig, D., Hausner, R., Wagner, W., Isaksson, C. (2006). Thermo-chemical storage for solar space heating in a single-family house. *Conference on Thermal Energy Storage*, Atlantic City, NJ.

Johannes, K., Kuznik, F., Hubert, J.L., Durier, F., Obrecht, C. (2015). Design and characterisation of a high powered energy dense zeolite thermal energy storage system for buildings. *Applied Energy*, 159, 80–86. doi: 10.1016/j.apenergy.2015.08.109.

Köll, R., van Helden, W., Engel, G., Wagner, W., Dang, B., Jänchen, J., Kerskes, H., Badenhop, T., Herzog, T. (2017). An experimental investigation of a realistic-scale seasonal solar adsorption storage system for buildings. *Solar Energy*, 155, 388–397. doi: 10.1016/j.solener.2017. 06.043.

Kousksou, T., Bruel, P., Jamil, A., El Rhafiki, T., Zeraouli, Y. (2014). Energy storage: Applications and challenges. *Solar Energy Materials and Solar Cells*, 120, 59–80. doi: 10.1016/j.solmat.2013.08.015.

Krönauer, A., Lävemann, E., Brückner, S., Hauer, A. (2015). Mobile sorption heat storage in industrial waste heat recovery. In *Energy Procedia,* 73, 272–280. doi: 10.1016/j.egypro.2015.07.688.

Kunii, D., Chisaki, T., Levenspiel, O. (1998). Novel rotary gas/solid contactor. *Powder Technology*, 96, 1–5.

Kuznik, F. and Johannes, K. (2020). Thermodynamic efficiency of water vapor/solid chemical sorption heat storage for buildings: Theoretical limits and integration considerations. *Applied Sciences*, 10, 489. doi: 10.3390/app10020489.

Kuznik, F., Johannes, K., Obrecht, C. (2015). Chemisorption heat storage in buildings: State-of-the-art and outlook. Energy and buildings. *SI: IEA-ECES Annex 31 Special Issue on Thermal Energy Storage*, 106, 183–191. doi: 10.1016/j.enbuild.2015.07.002.

Kuznik, F., Johannes, K., Obrecht, C., David, D. (2018). A review on recent developments in physisorption thermal energy storage for building applications. *Renewable and Sustainable Energy Reviews*, 94, 576–586. doi: 10.1016/j.rser.2018.06.038.

Lichtarowicz, M. (2013). Chemical reactors [Online]. Available at: https://www.essentialchemicalindustry.org/processes/chemical-reactors.html [Accessed 19 March 2021].

Lin, J., Zhao, Q., Huang, H., Mao, H., Liu, Y., Xiao, Y. (2021). Applications of low-temperature thermochemical energy storage systems for salt hydrates based on material classification: A review. *Solar Energy*, 214, 149–178. doi: 10.1016/j.solener.2020.11.055.

Makni, F. (2012). Développement d'un outil de simulation 2D-3D pour l'amélioration de la conception des absorbeurs dédiés aux systèmes de climatisation. PhD Thesis, Sorbonne Université, Paris.

Mauran, S., Lahmidi, H., Goetz, V. (2008). Solar heating and cooling by a thermochemical process. First experiments of a prototype storing 60 kWh by a solid/gas reaction. *Solar Energy*, 82, 623–636. doi: 10.1016/j.solener.2008.01.002.

Mette, B., Kerskes, H., Drück, H. (2012). Concepts of long-term thermochemical energy storage for solar thermal applications – Selected examples. In *Energy Procedia, 1st International Conference on Solar Heating and Cooling for Buildings and Industry (SHC 2012)*, 30(9), 321–330. doi: 10.1016/j.egypro.2012.11.038.

Mette, B., Kerskes, H., Drück, H., Müller-Steinhagen, H. (2013). New highly efficient regeneration process for thermochemical energy storage. *Applied Energy*, 109, 352–359. doi: 10.1016/j.apenergy.2013.01.087.

Michel, B., Mazet, N., Mauran, S., Stitou, D., Xu, J. (2012). Thermochemical process for seasonal storage of solar energy: Characterization and modeling of a high density reactive bed. *Energy, Asia-Pacific Forum on Renewable Energy*, 47, 553–563. doi: 10.1016/j.energy.2012.09.029.

Michel, B., Neveu, P., Mazet, N. (2014). Comparison of closed and open thermochemical processes, for long-term thermal energy storage applications. *Energy*, 72, 702–716. doi: 10.1016/j.energy.2014.05.097.

N'Tsoukpoe, K.E., Liu, H., Le Pierrès, N., Luo, L. (2009). A review on long-term sorption solar energy storage. *Renewable and Sustainable Energy Reviews*, 13, 2385–2396.

N'Tsoukpoe, K.E., Restuccia, G., Schmidt, T., Py, X. (2014). The size of sorbents in low pressure sorption or thermochemical energy storage processes. *Energy*, 77, 983–998. doi: 10.1016/j.energy.2014.10.013.

Nunez, T., Henning, H.M., Mittelbach, W. (2003). High energy density heat storage system – Achievements and future work. In *ISES 2003, Solar World Congress*, Göteborg.

Palomba, V., Vasta, S., Freni, A. (2017). Experimental testing of AQSOA FAM Z02/water adsorption system for heat and cold storage. *Applied Thermal Engineering*, 124, 967–974. doi: 10.1016/j.applthermaleng.2017.06.085.

Pinheiro, J.M., Valente, A.A., Salústio, S., Ferreira, N., Rocha, J., Silva, C.M. (2015). Application of the novel ETS-10/water pair in cyclic adsorption heating processes: Measurement of equilibrium and kinetics properties and simulation studies. *Applied Thermal Engineering*, 87, 412–423. doi: 10.1016/j.applthermaleng.2015.05.011.

Pinheiro, J.M., Salústio, S., Rocha, J., Valente, A.A., Silva, C.M. (2020). Adsorption heat pumps for heating applications. *Renewable and Sustainable Energy Reviews*, 119, 109528. doi: 10.1016/j.rser.2019.109528.

Rosen, M.A. (2001). The exergy of stratified thermal energy storages. *Solar Energy*, 71, 173–185. doi: 10.1016/S0038-092X(01)00036-6.

Scapino, L., Zondag, H.A., Van Bael, J., Diriken, J., Rindt, C.C.M. (2017). Sorption heat storage for long-term low-temperature applications: A review on the advancements at material and prototype scale. *Applied Energy*, 190, 920–948. doi: 10.1016/j.apenergy.2016.12.148.

Schultz, J.M., Andersen, E., Furbo, S. (2008). Advanced storage concepts for solar and low energy buildings. Final Report, IEA-SHC Task 32, Danmarks Tekniske Universitet (DTU).

Solé, A., Martorell, I., Cabeza, L.F. (2015). State of the art on gas–solid thermochemical energy storage systems and reactors for building applications. *Renewable and Sustainable Energy Reviews*, 47, 386–398. doi: 10.1016/j.rser.2015.03.077.

Srivastava, N.C. and Eames, I.W. (1998). A review of adsorbents and adsorbates in solid–vapour adsorption heat pump systems. *Applied Thermal Engineering*, 18, 707–714. doi: 10.1016/S1359-4311(97)00106-3.

Ströhle, S., Haselbacher, A., Jovanovic, Z.R., Steinfeld, A. (2016). The effect of the gas–solid contacting pattern in a high-temperature thermochemical energy storage on the performance of a concentrated solar power plant. *Energy Environ. Sci.*, 9, 1375–1389. doi: 10.1039/C5EE03204K.

Sun, L.M., Meunier, F., Brodu, N., Manero, M.H. (2016). Adsorption – Aspects théoriques. Opérations unitaires. *Génie de la réaction chimique* [Online]. Available at: https://www.techniques-ingenieur.fr/base-documentaire/42332210-opera-tions-unitaires-extractions-fluide-fluide-et-fluide-solide/download/j2730/adsorption.html [Accessed 20 April 2021].

Tatsidjodoung, P., Le Pierrès, N., Heintz, J., Lagre, D., Luo, L., Durier, F. (2016). Experimental and numerical investigations of a zeolite 13X/water reactor for solar heat storage in buildings. *Energy Conversion and Management*, 108, 488–500. doi: 10.1016/j.enconman.2015.11.011.

Thullner, K. (2010). Low-energy buildings in Europe – Standards, criteria and consequences. A study of nine European countries. Report, TVIT-5019.

Tulus, V., Boer, D., Cabeza, L.F., Jiménez, L., Guillén-Gosálbez, G. (2016). Enhanced thermal energy supply via central solar heating plants with seasonal storage: A multi-objective optimization approach. *Applied Energy*, 181, 549–561.

Weber, R., Asenbeck, S., Kerskes, H., Drück, H. (2016). SolSpaces – Testing and performance analysis of a segmented sorption store for solar thermal space heating. *Energy Procedia*, 91, 250–258. doi: 10. 1016/j.egypro.2016.06.214.

Xu, S.Z., Lemington, T., Wang, R.Z., Wang, L.W., Zhu, J. (2018). A zeolite 13X/magnesium sulfate–water sorption thermal energy storage device for domestic heating. *Energy Conversion and Management*, 171, 98–109. doi: 10.1016/j.enconman.2018.05.077.

Yu, N., Wang, R.Z., Wang, L.W. (2013). Sorption thermal storage for solar energy. *Progress in Energy and Combustion Science*, 39, 489–514. doi: 10.1016/j.pecs.2013.05.004.

Zettl, B., Englmair, G., Steinmaurer, G. (2014). Development of a revolving drum reactor for open-sorption heat storage processes. *Applied Thermal Engineering*, 70, 42–49. doi: 10.1016/j.applthermaleng.2014.04.069.

Zhang, Y. and Wang, R. (2020). Sorption thermal energy storage: Concept, process, applications and perspectives. *Energy Storage Materials*, 27, 352–369. doi: 10.1016/j.ensm.2020.02.024.

Zondag, H.A. (2015). Sorption heat storage. In *Solar Energy Storage*, Sørensen, B. (ed.). Academic Press, Boston. doi: 10.1016/B978-0-12-409540-3.00006-2.

Zondag, H.A., Van Essen, V.M., Schuitema, R., Bleijendaal, L.P.J., Kalbasenka, A.N., Van Helden, W.G.J., Bakker, M. (2009). Engineering assessment of reactor designs for thermochemical storage of solar heat. Technical Report, Energy Research Centre of the Netherlands.

4

Heat Storage by Chemical Sorption Processes

Antoine PERRIGOT, Driss STITOU and Maxime PERIER-MUZET
PROMES, CNRS, UPVD, Perpignan, France

4.1. Introduction

Sorption is a physicochemical phenomenon in which a gaseous chemical species, the sorbate, becomes bound to another material known as the sorbent, which can be a solid or liquid. Absorption refers to the process in which the sorbate penetrates into the volume of the sorbent, while adsorption refers to when the sorbate attaches to the surface of the sorbent. The large amount of heat involved in these phenomena makes them noteworthy for heat and/or cold storage applications.

Chemical sorption, or chemisorption, is a subcategory of adsorption. The adsorption phenomena may be classified according to the type of bond formed between the sorbent and the sorbate. This leaves us with two main families of adsorption: physical adsorption (or physisorption) and chemical adsorption (or chemisorption). In the case of physisorption, as described in Chapter 3, weak physical bonds, electrostatic in nature, are formed between the liquid sorbent and the gaseous sorbate. These can result from a hydrogen-type bond, or even from a Van der Walls-type interaction, for example. In the case of chemical sorption, the bonds are chemical, and thus we see covalent bonds, ionic bonds or even metallic bonds. Since chemical bonds are inherently stronger than physical bonds, the reaction enthalpies involved in chemisorption reactions are greater than in physisorption

Heat and Cold Storage 2,
coordinated by Nolwenn LE PIERRÈS and Lingai LUO.

reactions. This phenomenon gives chemisorption a potentially higher storage density.

Various types of chemical reactions of different natures can be carried out depending on the solid and gaseous reagents used. These reactions include complexation, coordination, hydrogenation and oxidation reactions.

The thermodynamic properties of chemical sorption reactions have made it possible to devise many different processes, such as storage, the accelerated production of heat or cold, or more recently, mechanical work. All such applications will be discussed later in this chapter, but particular attention will be paid to thermal storage processes using chemical sorption.

For the rest of this chapter, we will use the terms "chemical adsorption", "chemical sorption", "chemisorption" and "thermochemical system" interchangeably.

This chapter will be divided into the following sections. First, we present a brief history of thermochemical systems. Next, section 4.3 will describe the phenomenon of chemical sorption, from the level of the reactive material to the level of the process. Then, the state of the art of the various improvements in thermochemical processes is presented in section 4.4, while section 4.5 looks at the various applications of these chemical sorption processes that implement an intrinsic storage function.

4.2. History of chemical sorption systems

The first references in scientific literature to any efforts to make use of chemical sorption date back to the beginning of the 19th century, made by luminaries such as De Saussure and Faraday, among others. Faraday (1823) wanted to carry out an experiment to produce ammonia in its liquid state. He then used a salt reagent made up of silver chloride to adsorb ammonia gas. Then, using a heat source, he was able to release the ammonia by heating the reactive mixture, and then to liquefy the ammonia using a condenser. Faraday then observed the chemical reabsorption of ammonia by the salt, noting:

> Once the silver chloride is cooled, the ammonia immediately recombines with it, reproducing the original mixture. During this recomposition, a curious combination of effects takes place: when the chloride absorbs the ammonia, heat is produced, and

> the temperature rises to almost 100°C; while a few centimeters away on the other side of the tube, a considerable amount of cold is produced by the evaporation of the fluid.

The experiment, on Faraday's purely theoretical basis, opened the door to the use of thermochemical systems for the production of cold temperatures. The 20th century saw the advent of the first refrigeration systems based on chemical sorption. They first hit the market in 1960 with, among others, the commercial sale of a chemical sorption machine based on the reaction between ammonia and calcium chloride, devised by Plank and Kuprianoff. With the development of mechanical steam compression machines, research on sorption systems languished until regaining relevance in the 1970s, when it began to attract interest again thanks to spikes in energy prices. The first research done on the potential of using solar energy to operate such systems was carried out by figures such as Chernev, Meunier and Alefeld. Adsorption systems would be once again neglected by the scientific community for several more years, as absorption systems attracted greater interest instead. It was only by the second half of the 1980s, with the ban on the use of hydrochlorofluorocarbons (HCFCs) and chlorofluorocarbons (CFCs) as refrigerant fluids, that active research on solid sorption systems resumed; the period saw hundreds of new scientific articles published. This period also saw the first conferences to be held that focused on these systems. Today, research is mainly focused on improving the performance of chemical sorption systems, especially in terms of their coefficient of performance (COP) and their specific cooling power, which still remain two of the largest obstacles to industrializing these processes. Readers interested in a more in-depth examination of the history of chemical sorption systems may refer to articles by Dabrowski (2001) and Critoph (2012).

4.3. Principles of the operation of thermochemical systems

In order to provide an overarching explanation of how thermochemical systems work, we must first understand the phenomena involved on a smaller scale. Therefore, we will begin with a description of the thermodynamic properties of sorption phenomena, in particular those relating to their thermodynamic equilibrium and their chemical kinetics. Section 4.3.2 addresses the reactor, the component in which the reactive material is implemented and where the chemical reaction takes place. Different types of reactors are discussed, as well as different approaches to modeling them. Finally, we will provide a description of the operation of a "classic" thermochemical storage method implementing this type of chemical sorption reactor, with the goal of familiarizing readers with this type of system.

4.3.1. *The phenomenon of chemical sorption and reagents*

4.3.1.1. *Description of the reaction*

A chemical sorption reaction can be given the form of equation [4.1]:

$$S_0 + \nu.G \underset{\text{Decomposition}}{\overset{\text{Synthesis}}{\rightleftarrows}} S_1 + \nu.\Delta H_r \qquad [4.1]$$

In this equation, S_0 represents the sorbent before reacting with the gaseous sorbate G; S_0 is sometimes called a "poor" salt. The addition of a sorbate with gaseous stoichiometric proportions (ν) allows the salt S_1 to be formed, sometimes called a "rich salt". This reaction in which the gas is adsorbed by the salt is exothermal, releasing the enthalpy of reaction ΔH_r, which consists of the heat generated by the reaction involving one mole of gas. This direction of the reaction (from left to right) is called a "synthesis", referring to the rich salt, which is synthesized during this reaction. The reverse reaction, which is endothermic, is called a "decomposition" since the rich salt S_1 decomposes into poor salt S_0 through the effects of a supply of heat, later releasing the gaseous sorbate.

An example of a chemical sorption reaction can be seen in equation [4.2], which shows the reaction between barium chloride ($BaCl_2$) and ammonia (NH_3):

$$BaCl_2 + 8NH_3 \underset{\text{Decomposition}}{\overset{\text{Synthesis}}{\rightleftarrows}} BaCl_2(NH_3)_8 + 8\Delta H_r \qquad [4.2]$$

In the example above, the enthalpy of the reaction ΔH_r is equal to about 38 $kJ.mol^{-1}$ of gas, which means that a synthesis of one mole of the octa-ammionate $BaCl_2(NH_3)_8$ gives off about 300 kJ of heat due to the stoichiometric coefficient involved.

It is sometimes possible to then form several different salts from the same salt reagent for different operating conditions. For example, during the reaction between barium bromide and ammonia, the following by-products can be formed: $BaBr_2(NH_3)$, $BaBr_2(NH_3)_2$, $BaBr_2(NH_3)_4$, and $BaBr_2(NH_3)_8$. In order to correctly identify the reaction that takes place, a specific notation will be used: the term $BaBr_2(8/4)$ will refer to the synthesis reaction of $BaBr_2(NH_3)_8$ starting from $BaBr_2(NH_3)_4$.

4.3.1.2. *The gaseous sorbate*

The two most commonly used fluids in chemical sorption systems are water and ammonia. The most important properties of these are given in Table 4.1.

	Fluid	
	Ammonia	**Water**
Chemical formula	NH_3	H_2O
Boiling temperature (1 bar)	– 33.6°C	100°C
Melting temperature (1 bar)	– 77.7°C	0.00°C
Critical point	132°C | 113 bar	374°C | 221 bar
Molar mass	17.0 $g.mol^{-1}$	18.0 $g.mol^{-1}$

Table 4.1. *Properties of working fluids that are typically used in chemical sorption processes (source: data obtained with CoolProp)*

Other fluids, such as hydrogen, ethanol, methanol, carbon dioxide, methylamine or even dimethylamine, have also been studied in the scientific literature, but they do not appear as frequently.

The choice of a fluid is made on the basis of several criteria, as mentioned in Chapter 1 of this book. In particular, special attention must be paid to ensure that the fluid that is chosen is appropriate for the intended application. For example, it is important to make sure the working fluid never approaches the melting point; otherwise, the fluid will change to its solid state.

4.3.1.3. *The solid sorbent*

A large number of reactive salts can be found in the literature. The review made by Wang and Oliveira (2009) presents the state of the art of these reactive pairings. Sorbents can be grouped into three main categories: metal halides, hydrides and metal oxides.

4.3.1.3.1. Metal halide

Metal halides are chemical compounds consisting of a halogen (Cl, Br and I) and a metal, generally an alkaline-earth type metal (Mg, Ca, Sr, Ba) or a transition metal (Mn, Fe, Co, Ni). The chemical reactions with these salts are similar to complexation reactions, which generally have a high reaction enthalpy. These reactive pairs face issues of agglomeration and sintering, which require very particular attention to be paid as they are implemented in order to avoid these phenomena, which limit their reaction kinetics. The wide range of reactive couples available make it possible to use heat at temperature levels of up to approximately 300°C for ammonia and 500°C for water; beyond this, the decomposition of the reactive gas can occur, resulting from a catalytic action of the stainless steel used for the casing of the reactors.

4.3.1.3.2. Metal hydride

Hydrogen, which is used as a gaseous sorbate here, can react with many compounds with different reactions. For instance, it can react with lithium, calcium or even metal alloys to form compounds such as LiH or CaH_2. The use of hydrogen as sorbate implies the use of a system in which the sorbate will always be kept in the state of a gas, given that hydrogen is difficult to liquefy. Then, the hydrogen released during decomposition in a reactor will generally be captured to be synthesized in a second reactor. This is the principle of resorption systems, which will be described later in this chapter. The reactors of this type of system are often very heavy, but not very bulky. Therefore, this type of reactor containing metal hydrides could be used in applications such as for thermal storage in buildings where space is a much more significant constraint than weight. An evaluation of systems based on this type of salt for heat pumps and cold production applications is presented in Willers and Groll (1999).

This type of salt makes it possible to achieve a greater temperature increase than with metal halides (up to 400°C.). In addition, these salts often have the advantage of having high reaction kinetics. However, the thermal conductivity of these salts tends to decrease over time.

4.3.1.3.3. Metal oxide

Metal oxides can also act as sorbents, since they can take part in sorption reactions with oxygen, water, or even carbon dioxide. Calcium, lead, magnesium, and titanium oxides have been examined in the literature.

Although these salts are only used with heat sources that have a temperature above 150°C, they nevertheless make it possible for a greater increase in temperature than the two previous categories of salts (up to 1,000°C). The reaction enthalpy of these salts is generally high.

4.3.1.4. *Thermodynamic equilibrium*

Contrary to the physical absorption and adsorption phenomena described in Chapters 2 and 3, the thermodynamic equilibrium associated with chemical sorption reactions is monovariant. Therefore, it is possible to use a single equation giving a relationship between the pressure and temperature at equilibrium for a chemical sorption reaction. A good approximation that is widely used in the scientific literature consists of describing the thermodynamic equilibrium of such reactions using the Clausius–Clapeyron relation:

$$P_{eq} = P_{ref}.\exp\left(\frac{\Delta_r S^\circ}{R} - \frac{\Delta_r H^\circ}{R.T_{eq}}\right) \quad [4.3]$$

This equation gives the equilibrium pressure of the reaction P_{eq} at the equilibrium temperature T_{eq}. $\Delta_r H^\circ$ and $\Delta_r S^\circ$ represent the enthalpy and the end of the reaction under standard conditions (P_{ref} = 1bar and T_{ref} = 25°C), respectively.

However, the use of this equation requires three assumptions:

– the gas is considered as an ideal gas;

– the heat of reaction is assumed to be constant and equals $\Delta_r H^\circ$ in the pressure/temperature range usually considered to be (1–50 bar; 0–500°C);

– the molar volume of the solid phases (reactive salts) is negligible compared to that of the gaseous phases.

By definition, it is necessary for the reactive system to be limited to moving out of its equilibrium in order for either a decomposition reaction or a synthesis reaction to be carried out. The monovariant nature of the system implies that if an intensive thermodynamic parameter is fixed (in terms of temperature or pressure, for example), the second parameter will adjust in such a way as to return the reactive system to a state of equilibrium. Thus, if it is desired for the system to remain out of equilibrium, it is then necessary to set a fixed operating temperature and pressure for the system in order to maintain a certain thermodynamic deviation with respect to the chemical equilibrium. Figure 4.1 shows an illustration of these deviations at equilibrium in a Clausius–Clapeyron diagram. This type of diagram is widely used to illustrate sorption phenomena, since it allows the equilibria to be represented by straight lines. If only the pressure P_c is fixed, the system will then be at thermodynamic equilibrium at the temperature $T_{eq}(P_c)$. If a stress temperature (T_{dec} in this case) that is higher than this equilibrium temperature is set for the reagent, the system is then thrown out of equilibrium, and a decomposition reaction can occur if the reagent S_1 loaded with the gas is present. Thus, in a similar way, for any operating stress point (T_c, P_c) applied to the reactive system located above the equilibrium curve, a synthesis reaction will occur.

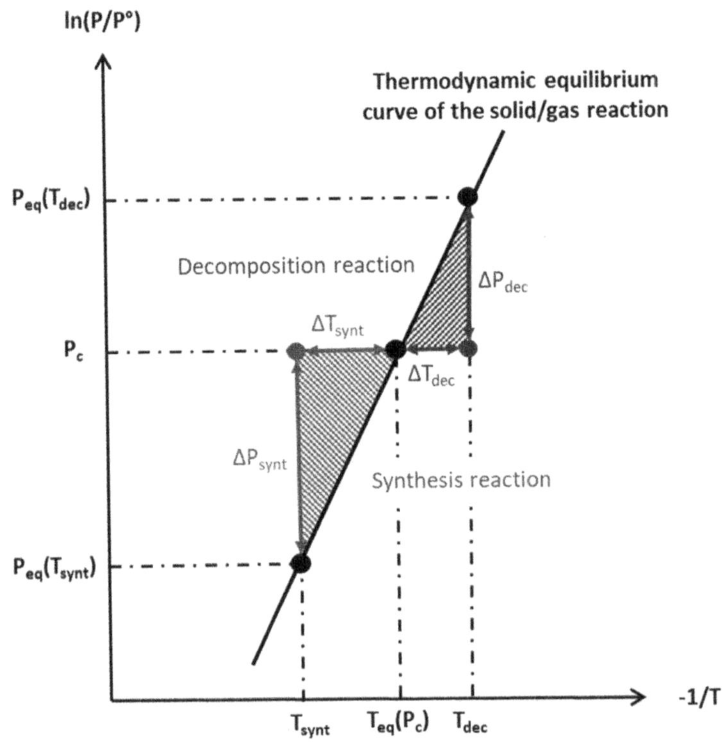

Figure 4.1. *Illustration in a Clausius–Clapeyron diagram of the deviations to the equilibrium that occur the reactive system in order to carry out a chemical synthesis or decomposition reaction*

4.3.1.5. *Chemical kinetics of the transformation*

The double constraint (T_c, P_c) applied to the reactor allows the reagent to be locally put out of equilibrium. Thus, in the most general case, the pressure and temperature gradients are established in the reagent block, whose shape will depend on the characteristics of mass and heat transfers.

We define the local progress X of the reaction as the proportion of synthesized salt (charged salt) relative to the total amount of salt present in a reactor; we therefore have:

$$\begin{cases} X = 1 & \text{if the synthesis reaction is completed} \\ X = 0 & \text{if the decomposition reaction is completed} \end{cases} \qquad [4.4]$$

We thus have a chemical kinetic that expresses the local rate of transformation of the reagent, and it is generally accepted that the more this kinetic increases, the greater the difference at equilibrium becomes. Several formulations have been proposed in the literature; an example from Mazet et al. (1991) is shown in equation [4.5]:

$$\frac{dX}{dt} = \begin{cases} K_{cin,d}.\,X.\left[\frac{P_{r,eq}-P_r}{P_{r,eq}}\right] & \text{for the decomposition reaction} \\ K_{cin,s}.\,(1-X).\left[\frac{P_{r,eq}-P_r}{P_{r,eq}}\right] & \text{for the synthesis reaction} \end{cases} \qquad [4.5]$$

In this equation, considering a reactor at uniform temperature and pressure, the term P_r refers to the actual pressure in the reactor, while the term $P_{r,eq}$ refers to the pressure that the reactor would be under if it were at equilibrium at the actual temperature of the reactor. The coefficients K_{cin} represent the kinetic constants of the reaction. They are generally on the order of 10^{-3} s^{-1} and can be obtained experimentally if a more precise value is desired.

4.3.2. *Thermochemical reactor*

The reactor is where the sorption reaction between the sorbate and the sorbent occurs. As illustrated in Figure 4.2, the reactor must both maximize the transport of material allowing the sorbent and the sorbate to come into contact and maximize heat transfers.

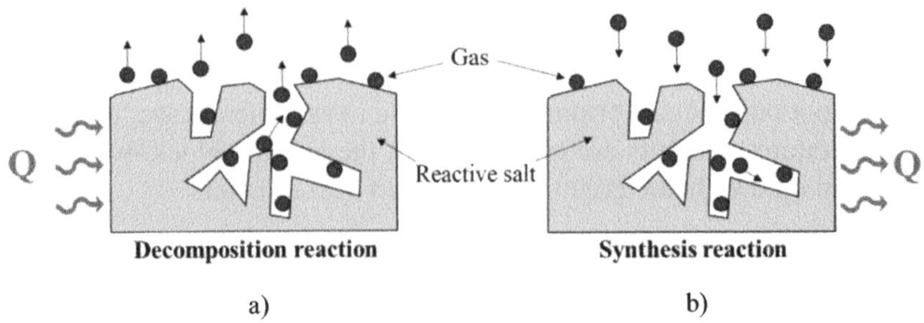

Figure 4.2. *Illustration of (a) a decomposition reaction and (b) a synthesis reaction*

4.3.2.1. *Typology of reactors*

Several typologies of reactors exist, as well as various methods for classifying them. The most commonly used of these methods consists of classifying the reactors according to the mode by which the reactive solid is transported. Thus, we can identify three main families of reactors:

– Stacked or fixed bed reactors contain a salt which can be fixed in the reactor, driven by a conveyor belt, stirred by rotation or even vibrated.

– Fluidized bed reactors, also called suspended bed reactors, are formed by the suspension of the reactive mixture through which the gas passes. In this way, the solid is present in the form of a powder, and the suspension is caused either by causing grains to fall due to gravity or by blowing the gas through the powder.

– Driven or vortex bed reactors are composed of a bed of particles transported by a fast-flowing gas.

Each of these three types is illustrated in Figure 4.3.

The use of only pure salt in a fixed bed reactor, though possible, often causes agglomeration problems that alter the transport of gas in the reactor over time. Effectively, the circulation of the reactive gas within the porous medium must not be a limiting factor for the progress of the reaction, particularly during synthesis reactions during which a significant increase in volume of the reactive solid is observed, decreasing the gas permeability. Moreover, the reaction powers are greatly limited by the very low thermal conductivity of the salt (on the order of 0.1–0.5 $W/m^2.K$). In order to promote the mass transfer in a reactor, it is common to mix the salt with a porous material, such as expanded natural graphite, vermiculite or even metal foam. As mentioned in Chapter 1, these materials are selected for their high thermal conductivity and their porosity, which contributes to improved diffusion of the gas within the reagent. These materials also make it possible to improve its mechanical properties, in particular the elasticity of the reactive mixture, which then makes it possible to absorb the variations in volume of the reagent during the reactions, without excessively limiting the quality of the mass transfers.

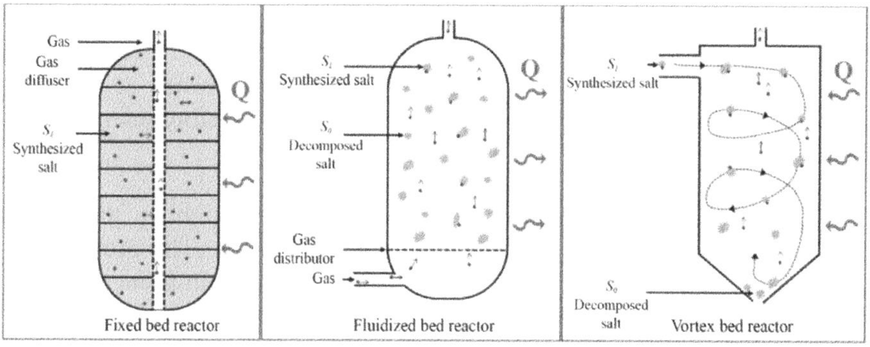

Figure 4.3. *Diagram of a stacked bed, a fluidized bed and a driven bed. For a color version of this figure, see www.iste.co.uk/lepierres/heat2.zip*

Nevertheless, it is important to note that the thermal and mass conductivities change in opposing directions during the addition of a porous material: while the material transport is improved during the addition of a porous material, the mass transport is reduced. It follows from this observation that there is an optimal amount to add to a pure salt in order to obtain the best possible performance.

4.3.2.2. *Modeling a reactor*

Many different models have been developed to describe the thermodynamic state of the reactor. The first models described in the literature come from the description of a reactive volume element, taking into account the transfers of matter and heat at the local scale. Then, a thermodynamic state of the system is defined locally, accounting for the pressure, the temperature and the progress of the reaction at each moment. This model is described by Lu et al. (1996), among others, based on the resolution of the differential equations deriving from local thermal and mass balances associated with local chemical kinetics. By solving these equations, it can be established that the overall transformation of a reactor is linked to the progression of two mostly diffuse reaction fronts based on the characteristic parameters of the transfers, one relating to the quality of diffusion of the reactive gas which affects the local pressure and the second to the heat diffusion which impacts the local temperature. The use of this type of model allows a fine understanding of the local phenomena taking place in the reactor, though it also requires knowledge of many of the characteristic parameters of the reactive medium. A few different reductions of this model have now been proposed.

First, a simplification of this model, described in Hui-Bo et al. (1997) consists of assuming that the chemical kinetics, defined locally as the speed of transformation of the reagent, are infinite. We can therefore refer to it as a model with steep fronts.

We can also assume that once the pressure is sufficient in the reactor ($P_c > 1$ bar), the mass transfer is no longer a limitation. The pressure can then be considered uniform in the reactor, and this results in a single thermal front model (Stitou et al. 1997).

A second method of modeling thermochemical reactors starts with the assumption that the pressure and the temperature are uniform within the reactor. These models, classified as nodal models, allow us to obtain a simplified representation of the transformation using the form of a system of differential equations that are simpler to solve. This type of nodal model is established by considering internal energy and mass balances at the scale of the reactive mixture as a whole. It is also necessary to integrate this type of model with an overall kinetic reaction time, taking into account the difference in the thermodynamic equilibrium applied to the reactive mixture. This type of model has been used in applications such as modeling the behavior of a cold production process (Le Pierrès et al. 2007a and 2007b).

Finally, another modeling approach, using a stationary-type model, can also be used for applications such as estimating the energy performance of a chemical sorption process. By relying solely on the thermodynamic properties of the reactions implemented (enthalpy and reaction entropy), it is possible to simply estimate the different energy transfers, and thus obtain outputs such as performance coefficient values. In this type of model, it is assumed that the temperature and the pressure are constant and uniform in the reactor as well as in the different exchangers with which the reactor interacts: this simplified approach is called steady-state modeling. This type of modeling makes it possible to perform operations such as evaluating the performances of a sorption system assisted by a compressor (Ferrucci et al. 2018).

4.3.3. *Typical method for thermochemical storage*

The phenomenon of chemical sorption and thermochemical reagents and reactors were described in the previous sections. To have an idea of how a heat storage system can be designed, the thermochemical system has to be studied at the scale of the process. The application of thermochemical sorption systems to heat storage is possible thanks to the reversibility of the sorption phenomena that are involved. The heat that is desired to be stored is used to carry out the desorption reaction, and thus separate the sorbate, in the form of a gas, from the solid sorbent material. The gas produced in this way by this decompression can be recovered and stored. During the discharge phase, the sorbent and the sorbate are brought back into contact, and the exothermic sorption reaction occurs, releasing the stored heat in chemical form, which can then be reused. These processes can be classified into two major families: open systems and closed systems. A system is said to be open when the sorbate is drawn in from and later discharged back into the surrounding medium, while in closed systems, the sorbate undergoes a cycle of transformation in the process.

In order to circulate the sorbate in a closed system, it is necessary to couple the sorption reaction with another physicochemical process. It is possible to use the changing of the sorbate's state between liquid and gas (the basis thermochemical system) or another sorption reaction; such systems are known as resorption systems.

4.3.3.1. *Closed system coupled with a liquid/gas state change*

The heat that is stored is mainly used to desorb the gas from the sorbent material, which is then condensed and stored at room temperature, generally in the form of a liquid. During the discharge phase, the liquid evaporates and is reabsorbed by the material. Ideally, if this evaporation is carried out at the same temperature as the condensation, the heat produced during this chemical reabsorption is delivered at the same temperature as the heat provided during the sorbent decomposition. This ideal storage is described in Figure 4.4.

Actually, it is impossible to restore the heat to the same temperature as during the storage phase because of the different temperature differences that are called thermal pinches. These temperature differences are necessary to carry out the heat transfers in heat exchangers, changes in the thermodynamic equilibrium necessary for the sorption reactions to take place or compensate pressure losses in the ducts.

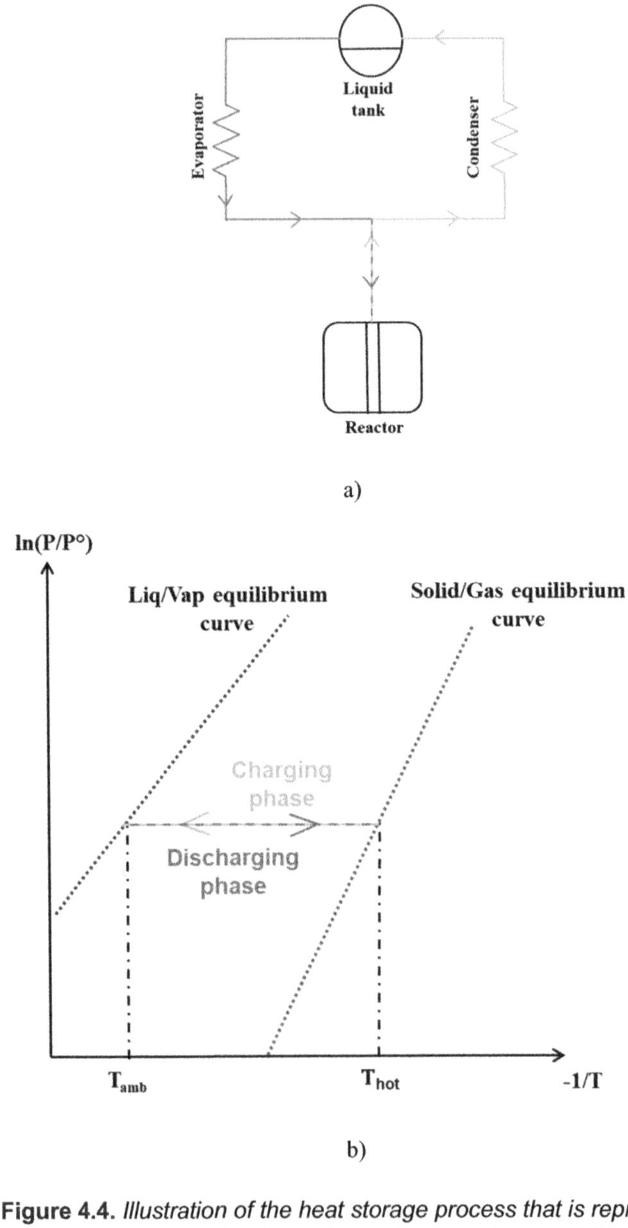

Figure 4.4. *Illustration of the heat storage process that is represented operating ideally under thermodynamic equilibrium conditions. For a color version of this figure, see www.iste.co.uk/lepierres/heat2.zip*

As noted at the beginning of this chapter, it is possible to obtain various useful energy effects from the stored heat. One useful effect widely studied in the scientific literature is the production of cold. The most common cycle for the production of cold is presented below using several thermodynamic diagrams.

In order to simplify these diagrams, the deviations at equilibrium which allow the different reactions or heat exchanges to take place are not represented here: the operating conditions are those of the thermodynamic equilibria.

The cycle operates according to two discontinuous phases represented in green (charging phase, decomposition of the salt reagent) and purple (discharge or cold production phase).

Each point shown in Figure 4.5 represents the thermodynamic state of the work fluid at a given point in time.

We can therefore describe the cycle as follows:

– ⓐ → ⓑ, the thermal energy is supplied to the reactor in order to raise the temperature and the pressure, and so it is suitable for the storage phase;

– ⓑ → ⓒ, when the pressure in the reactor reaches the prevailing pressure in the condenser, these two components are connected so that the thermochemical storage phase can be completed. The condenser then changes the reactor's pressure to its own pressure level. The pressure is therefore set by the condenser while the temperature is set by the heat supply; the system is now taken out of equilibrium, while the gas is desorbed and flows toward the condenser;

– ⓒ → ⓓ, the gas cools down until it reaches the state of a saturating vapor (preventing the vapors from overheating);

– ⓓ → ⓔ, the gas condenses until the work fluid reaches the state of a saturated liquid;

– ⓔ → ⓕ, the liquid is pulled by gravity into the tank, where it continues to cool down to room temperature. The liquid stored in this way no longer undergoes thermal losses, and it is now possible to carry out long-term storage if so desired by disconnecting the reactor of the condenser;

– ⓕ → ⓖ, when the user wishes to draw from storage and the reactor is returned to room temperature, the synthesis phase (or production of cold) can begin. The liquid then expands to the vaporization pressure;

– ⓖ → ⓗ, the fluid vaporizes until reaching the state of a saturated vapor. Here, we may observe a possible useful effect of the cycle, in that the endothermic character of evaporation offers the possibility to produce cold;

– ⓗ → ⓘ, the gas produced is sucked in by the reactor and warms up as it comes into contact with the hotter reagent;

– ⓘ → ⓙ, the gas is chemically absorbed by the reactive salt, releasing heat. This heat can also be reused if necessary;

– ⓙ → ⓚ, once all the liquid has been vaporized, the sorption reaction is disrupted and the reactor returns to ambient temperature and to the corresponding equilibrium pressure for this temperature.

Although the Clausius–Clapeyron diagram has the advantage of representing the equilibria (pressure/temperature) of the sorption processes simply by using straight lines, certain operating points can be superimposed, and it is not always easy to grasp the concrete functioning of the system.

The enthalpy diagrams (or Mollier diagrams) and entropy diagrams (or T-s diagrams) provide us with a way of getting past this obstacle. In addition, they make it possible to better understand the flows of energies and exergies within the system since the values of enthalpies and entropies are presented in these diagrams.

Figures 4.6 and 4.7 give a representation of the thermodynamic cycle presented previously in a Mollier diagram and in a T-S diagram.

The Mollier diagram (Figure 4.6) is useful for estimating the different amounts of energy exchanged during the thermodynamic cycle.

For example, we can estimate the thermal energy released by the condensation of 1 kg of ammonia by considering the difference in enthalpy between the points ⓓ and ⓔ, which in this case is 1.1 MJ or 0.3 kWh. If the different amounts of energy are known, it is then possible to estimate the performance of the system.

The temperature–entropy diagram (Figure 4.7) provides additional information about the heat fluxes within the system. The heat consumed in the case of any transformation is equal to the area between the curve describing this transformation and the horizontal line at T = 0 K.

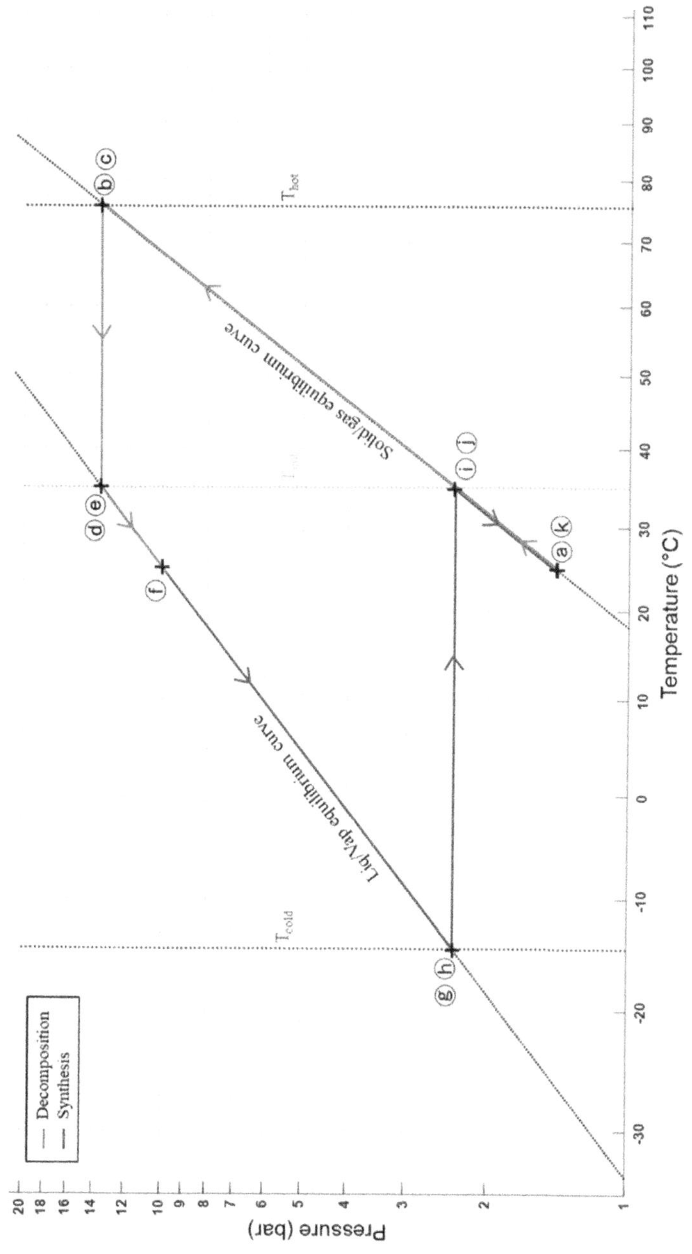

Figure 4.5. *Clausius–Clapeyron diagram of a classical sorption cycle allowing for the storage of heat for the purpose of a delayed production of cold. For a color version of this figure, see www.iste.co.uk/lepierres/heat2.zip*

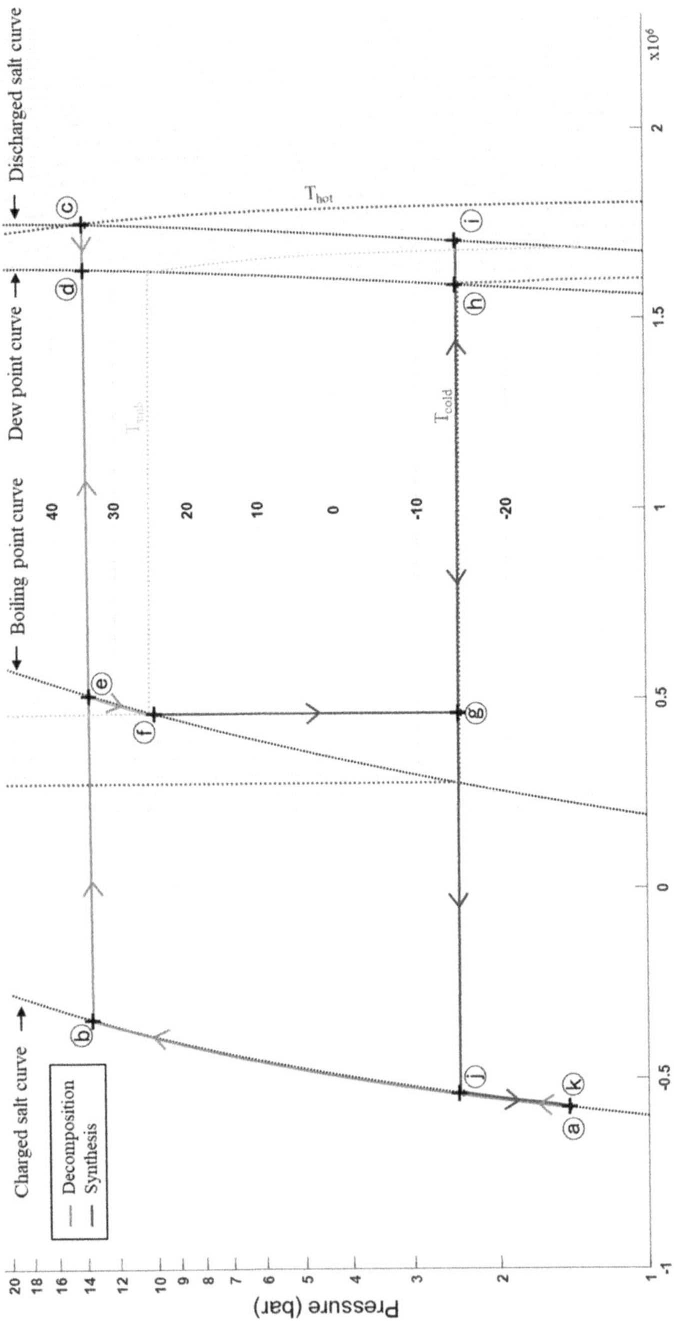

Figure 4.6. *Mollier diagram of a classical sorption cycle allowing for the storage of heat for the purpose of the delayed production of cold. For a color version of this figure, see www.iste.co.uk/lepierres/heat2.zip*

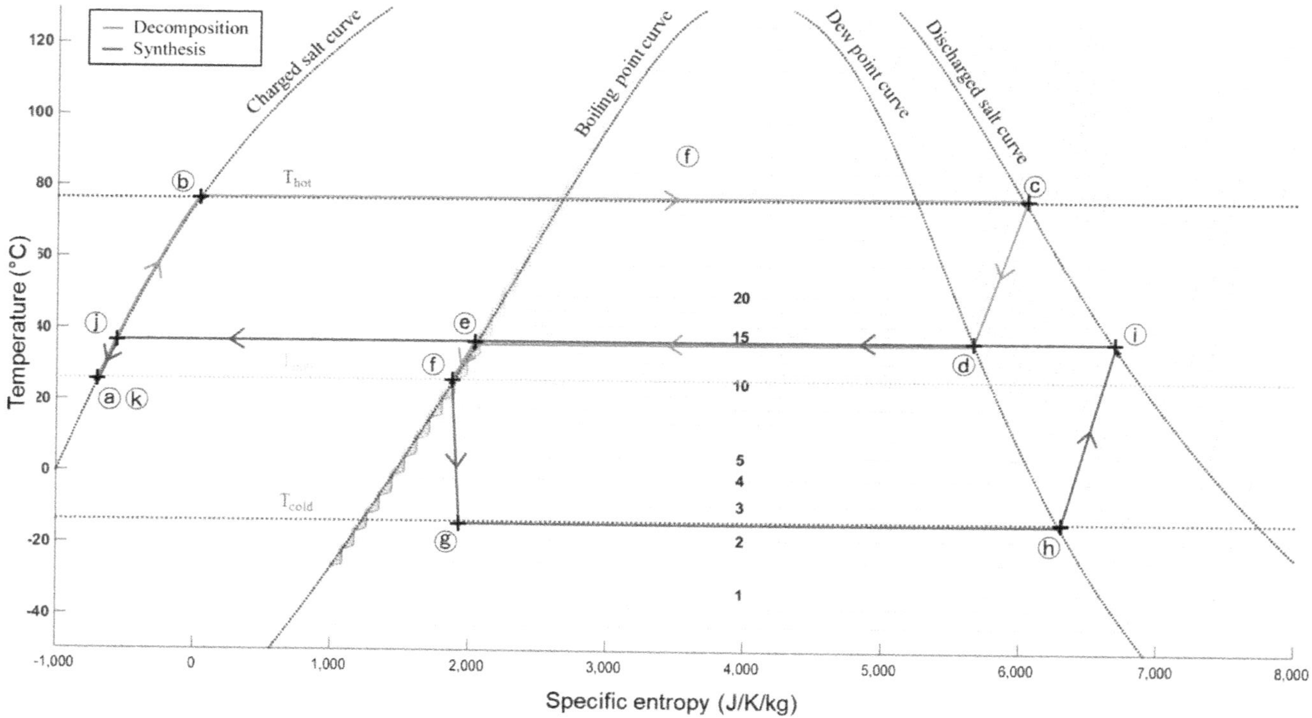

Figure 4.7. *Temperature–entropy diagram of a classical sorption cycle allowing for the storage of heat for the purpose of the delayed production of cold. For a color version of this figure, see www.iste.co.uk/lepierres/heat2.zip*

4.3.3.2. *Closed resorption system*

It is also possible to pair two sorption processes together to obtain a cycle similar to the previous one, bringing together a reversible chemical reaction and a state change process. In fact, the gas desorbed from the first reactor can also be absorbed by a different sorbent material placed in a second reactor.

The major advantage of such a pairing is that this can create a storage method that uses only solid active materials with the same gas, thus avoiding the need to manage liquid sorbate, which can be complex and create problems, especially in applications such as transportation.

An illustration of a thermochemical resorption cycle, as well as the "classic" cycle of cold production, is presented in Figure 4.8.

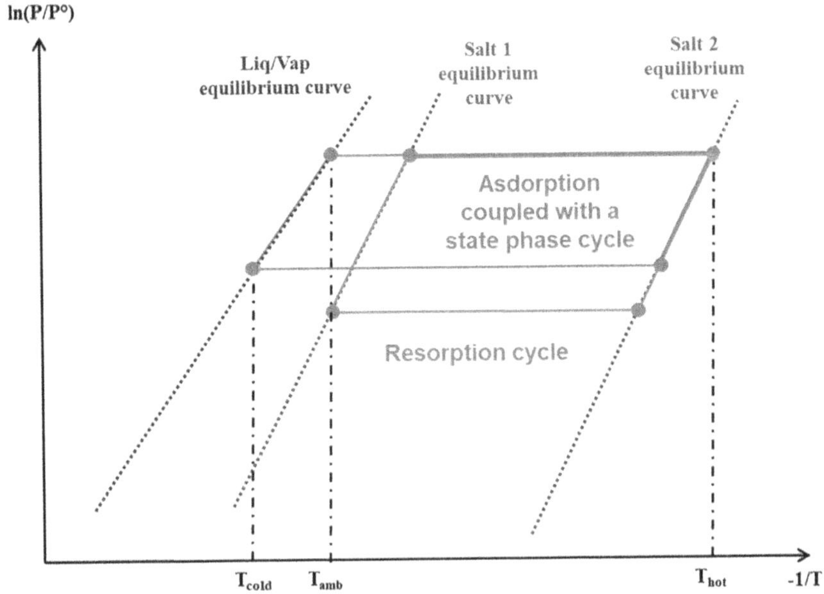

Figure 4.8. *Chemical sorption cycle for a coupling with a change of state and for resorption. For a color version of this figure, see www.iste.co.uk/lepierres/heat2.zip*

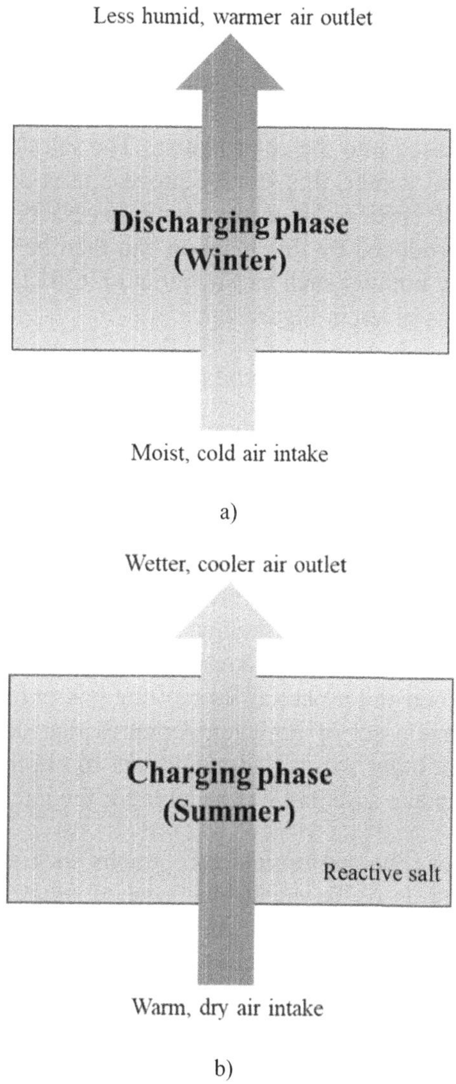

Figure 4.9. *Illustration of a chemical sorption process in an open system*

4.3.3.3. *Open system*

Open systems have the unique feature of taking in and discharging the gas necessary for the sorption reaction from the surrounding environment, usually the water vapor contained in the ambient air. This is why the majority of open systems use materials based on salt hydrates or hydroxides reacting with the humidity of the ambient air. A classic example involves storage in the summer; the desorption of the reactive mixture produces water vapor, which is released into the environment. The salt desorbed in this way is then stored. In the winter, this storage process uses the humid air in the environment, which carries the water vapor necessary for the synthesis reaction: the heat produced by this reaction can then be used to heat living spaces, as shown by authors such as Michel et al. (2012). An illustration of this type of method is given in Figure 4.9.

4.3.3.4. *Criteria for the choice of salts on the basis of the application*

The various different salts mentioned above in this chapter must be chosen carefully on the basis of their intended application. For example, it is necessary for the temperature of the heat source that we wish to use to be compatible with the temperature necessary for the decomposition of the reactive salt. In the case of a cold production system where the sorption reaction is paired with a change of state reaction, for example, the cold production cannot be carried out below a certain temperature, since it is necessary for the vaporization pressure to remain higher than the synthesis pressure in order to ensure a natural flow of the gas to the reactor. Because of this, each salt has a set of limit temperatures that define the operating range of the salt. In other words, it is important to observe the equilibrium curve of the salt to make sure that it is suitable for the intended application.

In order to choose the appropriate salt, graphs such as that presented in Figure 4.10 may be used. To create this type of figure, it is necessary to know the enthalpies and entropies of each reaction in order to determine the equilibrium curves of each reaction with the Clausius–Clapeyron equation. It is then possible to define the temperatures and the thermal pinches at the level of each heat exchange.

For example, if we consider that the condenser uses an exchange with a heat transfer fluid (such as water), a thermal pinch of the order of 5°C can be considered. If we also consider a thermal deviation for the equilibrium (which is necessary for the reaction to take place) on the order of 10°C, we obtain an overall pinch of 15°C. If the exchange is carried out with air, it will be necessary to adjust the value of the pinch, and therefore it would be more appropriate to assume an overall thermal pinch on the order of 20°C.

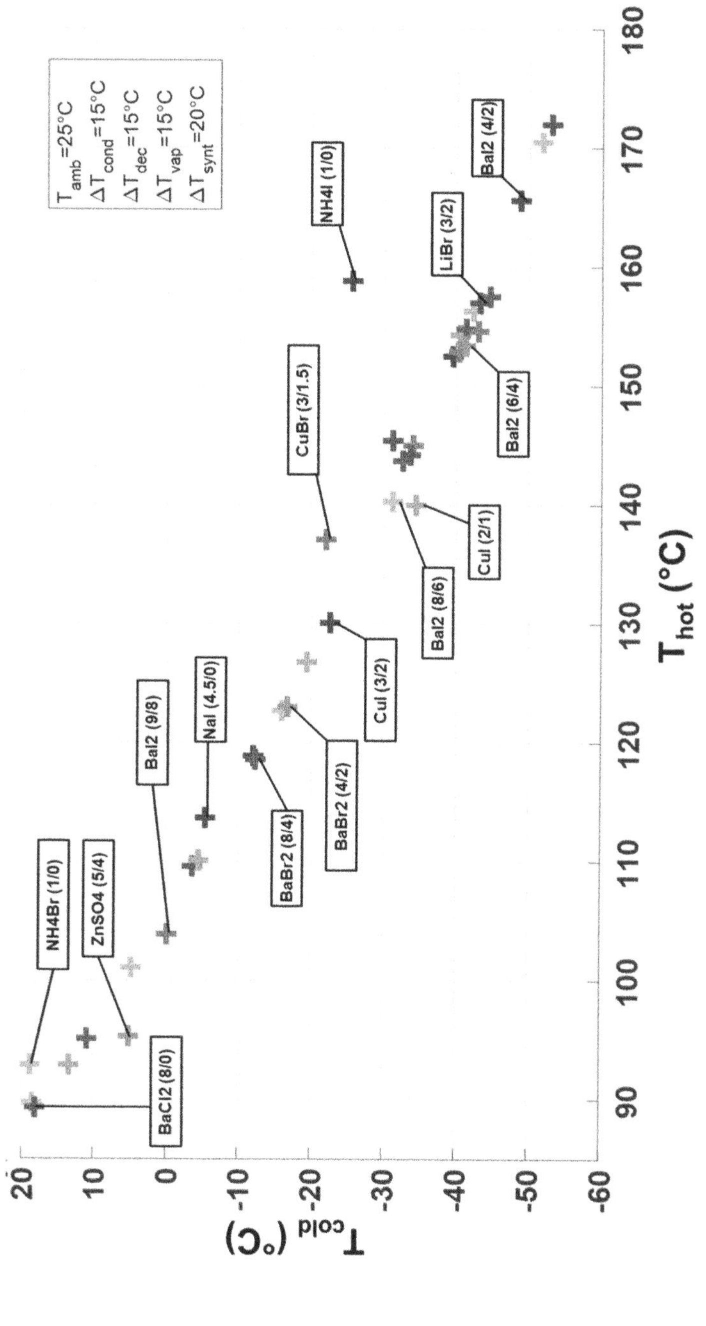

Figure 4.10. *Distribution of ammoniates on the basis of their limits of use, depending on the level of useful cold to be produced by the thermochemical system and the heat source available for its regeneration. For a color version of this figure, see www.iste.co.uk/lepierres/heat2.zip*

If the temperature of the thermal energy source that we want to store is already determined, then it is necessary to choose a salt that allows for a stress temperature that is satisfactory for the pressure of the reactor or higher than the condensation pressure.

For example, if we want to design a storage system to recover heat at 100°C, only the six ammoniates to the left of the vertical straight line $T = 100$°C can be used.

In addition to being suitable for the intended application, the salt must also offer the best possible compromise between the following constraints:

– The salt must present the lowest possible risk of harm (e.g. using lead requires special precautions).

– It is preferable to use a salt with the highest possible energy density.

– The temperatures reached during the cycle must not deteriorate the salt (through melting, crystallization, etc.).

– A process for shaping the salt must be used, which is as easy to implement as possible.

– The use of certain salts may be restricted by their availability on the market, since certain materials are considered to be scarce resources.

4.3.3.5. *Performance criteria for chemical sorption systems*

The energy performance of thermochemical processes can be evaluated using several different criteria. The indicators that appear to be the most relevant are the following:

– *the COP*, expressed as the ratio of the useful energy produced to the high-valuecostly energy consumed. If two COPs are to be compared with each other, it is nevertheless necessary to make sure that they have been calculated under the same conditions. The differences may be quite significant depending on the COP calculation method – for instance, a COP calculated from thermodynamic data is always higher than a COP obtained experimentally;

– *the storage energy density*, which translates the amount of stored energy reduced to a unit of volume or mass of the system. Nevertheless, it is necessary to pay close attention to what is being considered in the calculation of these densities, in particular to the calculation of the stored energy or the restored energy, as well as the unit of volume or mass consumed (whether it involves the active material, the reactor or the total system);

– *the specific power of thermal production*, which makes it possible to evaluate the thermal power of the process (refrigerating or heating) reduced to a unit of volume or mass of the system. The same issues as those affecting energy density should be kept in mind when this indicator is used.

4.4. Advanced thermochemical processes

Different process configurations and different thermodynamic cycles have been studied in the literature in order to improve the performance of these systems or to allow them to produce new useful effects. In particular, in many cases a second reactor is used, because this makes it possible to obtain continuous processes, which were discontinuous up to the time this method was introduced. For example, in the case of a process using a single fixed bed reactor, it is impossible to simultaneously charge and discharge the system, as would be done with an electrochemical battery in the case of storing electricity. The use of a second reactor thus makes it possible to have one reactor in the charging phase and the other in the discharging phase. The functions of the two reactors are then reversed. Nevertheless, it must be noted that at the time when the functions of the two reactors are swapped, it is necessary to raise the temperature and pressure of the reactor which was in synthesis, and vice versa for the second reactor. To minimize this energy loss, several thermodynamic cycles have been proposed, such as heat and mass recovery cycles.

4.4.1. *Heat recovery cycles*

In a classic cycle, the alternation between the decomposition and synthesis phases of a reactor generates a significant loss of sensitive heat, which makes it possible to place the reactor in thermodynamic conditions that are apt for the phase that is desired to be carried out. This portion of the heat, even though it is necessary, does not directly participate in sorption and can therefore be considered to be lost. When using a second reactor, during the time when the functions of each reactor switch, it is possible to carry out a thermal transfer from the hotter reactor to the colder one, in order to increase the colder reactor's temperature. An illustration of this method is given in Figure 4.11.

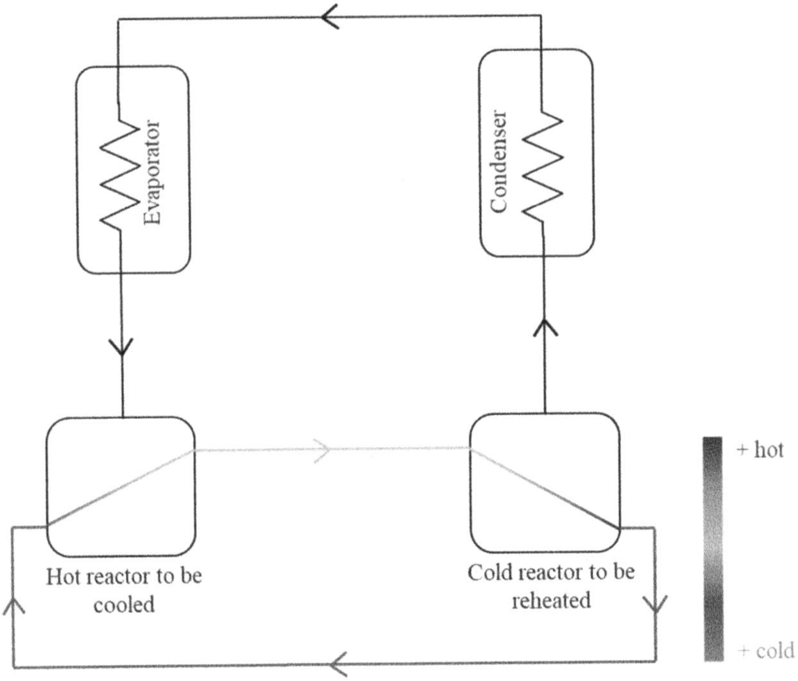

Figure 4.11. *Illustration of a closed heat recovery cycle. For a color version of this figure, see www.iste.co.uk/lepierres/heat2.zip*

The use of a heat carrying fluid for transferring the temperature from the hotter reactor to the cooler reactor can be improved by adding a system for heating and cooling the fluid (Pons 1997). In fact, it can be shown that a slight overheating of the heat transfer fluid before it reaches the "cold" reactor makes it possible to transmit a greater portion of the thermal energy to the reactor.

Similarly, by cooling the transfer fluid between the reactor undergoing decomposition and the one undergoing synthesis, it is possible to recover a larger amount of heat in the synthesis reactor (see Figure 4.12).

The same process has also been envisaged using the work fluid as a heat transfer fluid (Critoph 1994).

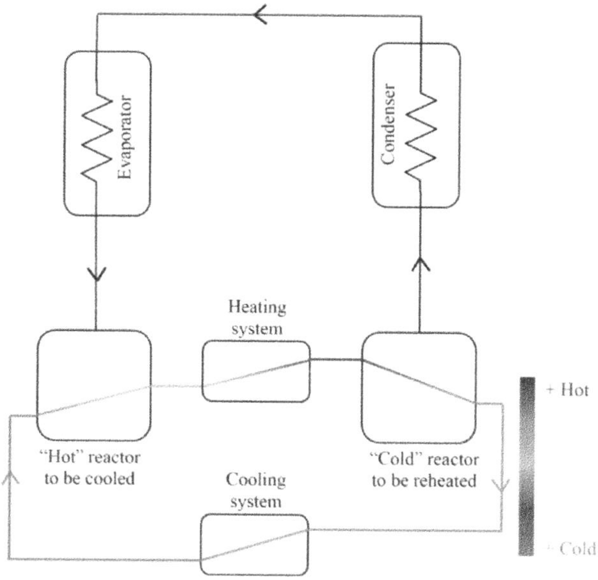

Figure 4.12. *Diagram of the enhanced heat recovery cycle with heat wave. For a color version of this figure, see www.iste.co.uk/lepierres/heat2.zip*

4.4.2. *Mass recovery cycles*

Another possibility for improving thermochemical processes using two reactors consists of physically linking the reactors at the time of the phase change (Lu et al. 2014). The pressure difference between the reactor at the end of decomposition (high pressure) and the one at the end of the synthesis (low pressure) will then generate a transfer of material until the pressure is equalized. This therefore promotes the lowering of the reactor pressure, which must meet the thermodynamic conditions of the synthesis while also promoting the rise in pressure of the reactor, which must meet the operating conditions of the decomposition.

4.4.3. *Multi-effect thermochemical processes in thermal cascades*

Another path for improving chemical sorption processes consists of coupling several thermodynamic cycles together, and then by using the thermal discharges of some of these as a driving source for others, we obtain multi-effect cycles or multi-stage cycles in thermal cascades that are more complex, but of increased efficiency.

The thermal cascade cycles, or cascaded cycles, allow for work pairings of differing natures to be implemented. There are many combinations found in the literature that can be done either between salts of the same type (such as ammoniates), for which the work fluid is the same, or combinations from different systems (absorption/adsorption, adsorption/resorption, etc.). Stitou et al. (2003) proposed, for example, to couple a chemical sorption system to an absorption system. The heat rejected by the first system is reused in the second system.

The use of multi-effect cycles makes it possible to take advantage of useful effects different from the classic cycles. Indeed, in addition to storage functions, we may also consider this for uses such as the production of low-temperature cold or high-temperature heat.

4.4.4. *Hybrid thermal/mechanical cycles*

4.4.4.1. *Hybrid cycles using a compressor*

When it is desired to store low-temperature thermal energy, it may be advantageous to pair the classic sorption system with a compressor. Essentially, if the temperature of the heat source to be stored is not sufficient to obtain the pressure necessary for the gas to flow, a mechanical compressor can fill this gap. It has even been demonstrated experimentally that it is possible to operate the system at ambient temperature. Articles such as (Ferrucci et al. 2018; Fitó et al. 2019) offer an example of the implementation of the compressor in a chemical sorption cycle to assist the decomposition phase.

Compressor-assisted synthesis is also a potential option under study at the PROMES laboratory. Using a compressor during the synthesis phase would make it possible to reduce the vaporization pressure, and thus to produce cold at a lower temperature. Thus, as in the case of multi-effect processes in thermal cascades, in addition to storage, it would be possible to produce low temperature cold.

Figure 4.13 shows the different configurations that can be obtained using a compressor.

4.4.4.2. *Hybrid cycle with work production*

Designs using a joint cold-temperature and electricity generation system have also been explored (Godefroy et al. 2019). This chapter proposes using a turbine to recover the mechanical energy available during the expansion of the fluid in a resorption cycle.

4.5. Diversification of applications with storage

The diverse range of salt/gas reactive pairings and thermodynamic cycles has made it possible to envisage a wide variety of applications. For the rest of this chapter, we will discuss some of the most notable applications that can be found in the scientific literature that include a storage function.

4.5.1. *Heating*

The most natural use of chemical sorption processes consists of reducing the heat released during the synthesis reaction. If the temperature of these thermal discharges is too low for the desired useful effect, cascaded cycles, implementing internal heat recovery may then be incorporated into the process. Yang et al. (2020) have presented a theoretical study on a heating system for housing in cold regions, where the outside temperature can drop below –25°C. This theoretical study was done on 121 ammoniates. The system, powered by a gas burner, could heat residential units with a COP of up to 1.22 for a single-effect cycle and 1.51 for a double effect cycle, taking into account the recovery of the condensation heat.

As detailed previously, the storage allowed by chemical sorption processes makes it possible to consider this type of system for long-term storage, since it does not have any natural discharge due to thermal losses. Li et al. (2013) have carried out a theoretical study on an inter-seasonal storage system. By collecting solar thermal energy in the summer, this system allows the reactive salt to be decomposed and the ammonia obtained through this process to be stored in liquid form in a tank at room temperature. During the winter, the synthesis process outputs the heat that is stored through this process. The study proposes to couple two reactors together to adapt to the level of the outdoor temperature and thus to be able to provide heat at a sufficiently high temperature, despite low outdoor temperatures. This system achieves a COP of approximately 0.6 and an energy density of more than 1,000 kJ per kilogram of salt.

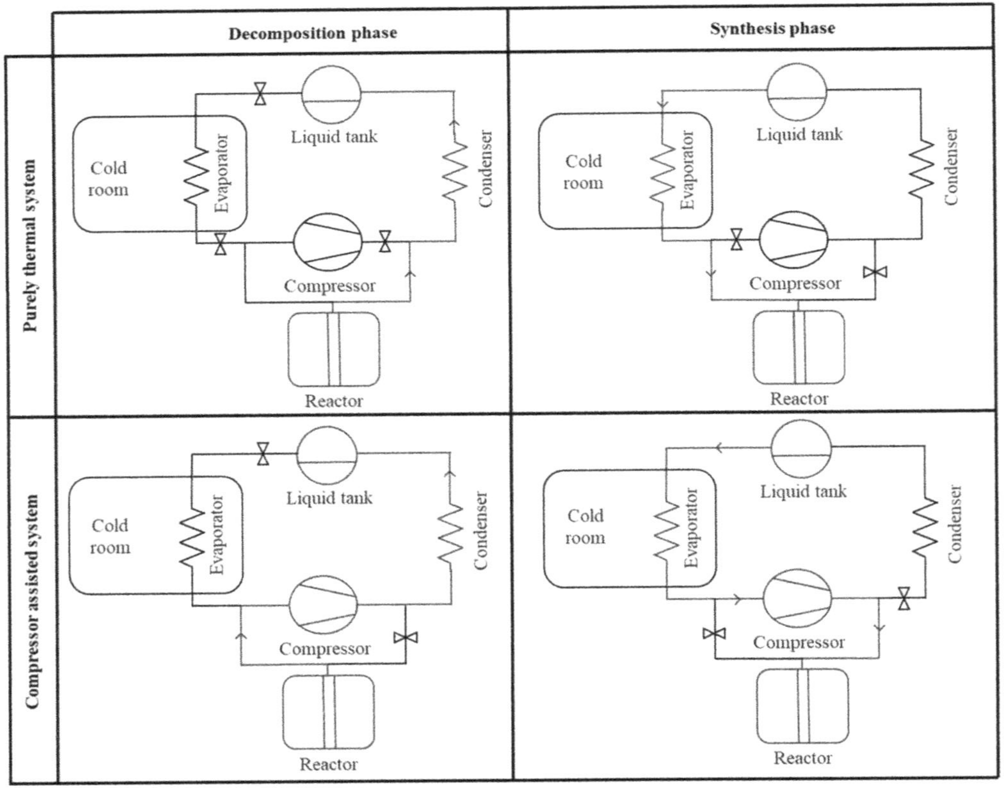

Figure 4.13. *Illustration of the different possible configurations for a hybrid cycle using a compressor*

Sorption processes have also been considered for high-temperature storage, particularly for the case of thermodynamic solar power plants (Boulnois 2019).

Drying food using solar energy is a common practice, especially in tropical and subtropical regions. Nevertheless, the use of a sorption system running on solar energy allows a significant improvement, because it allows for a finer level of control in managing the drying. Indeed, during the day, solar energy not only allows for drying food, but also for charging the thermochemical system. It is then possible to use this stored heat at night or on cloudy days. A prototype was tested in Malaysia (Fadhel et al. 2010) in which thermal discharges from condensation and synthesis were used for drying. The experiment showed that 85% of the energy required for drying could be supplied by the thermochemical system, with the rest being supplied by an auxiliary dryer. A COP of about 2 has been achieved for this system, which operates using solar energy and the chemical sorption reaction between the $CaCl_2$ and ammonia.

4.5.2. *Production of cold*

In a cycle that pairs together a sorption reaction with a change of state, the production of cold is done in the evaporator. The heat is supplied to the reactor to ensure the decomposition occurs. The other thermal exchanges take place with the ambient temperature. Stitou et al. (2012) present a prototype of solar air conditioning producing up to 20 kWh of cold (at 4°C) per day from 20 m^2 of thermal solar panels, providing heat at 60–70°C. This totals about 1 kWh of cold per m^2 of solar panels with an overall solar COP of around 20%. The use of such a system requires the decomposition phase of the salt reagent to be done during the day. Thus, solar thermal energy is stored and then used to produce cold during the night.

Resorption systems also offer interesting possibilities for the storage of cold, because the fluid is no longer stored in the liquid state. This means that the system can be moved more easily. In Li et al. (2010), a system is presented for producing cold at temperatures of 10°C from a heat source at 180°C using a resorption system. The system uses the pairing of $BaCl_2(8/0)$

and $MnCl_2$ (6/2). Another resorption system is presented in Bao et al. (2012). This system makes it possible to produce cold between −15°C and 5°C from a heat source between 140°C and 170°C. The experiment was done with the salts $MnCl_2$(6/2) and NH_4Cl (3/0).

By lowering the temperature of the cold produced, it is possible to use this for producing ice. Lu et al. (2007) uses a reactive mixture of $CaCl_2$ and activated carbon to produce ice from a heat source at 126°C. In particular, this article studies a heat and mass recovery cycle, and proposes a system for producing ice onboard a fishing boat. This system can potentially produce up to 17 kg of ice per hour.

The use of advanced cycles presented previously makes it possible to lower the temperature even lower. Le Pierrès et al. (2007a, 2007b) present, in particular, the theoretical study and the experimental results for the production of very low temperature cold. The experiment showed how it would be possible to cool a 560 L cold chamber at −25°C using the heat produced by two flat thermal solar panels producing heat at 70°C. To achieve this, a cascade cycle composed of two reactors using the pairing $BaCl_2/NH_3$ was used. The storage function included in this system carried out a cold production that is delayed after the period of sunlight intake.

4.5.3. *Simultaneous cold/heat/work generation*

Some applications propose pairing together the two previous useful effects given in Lahmidi et al. (2006) and Mauran et al. (2008), which present a prototype for thermal storage that produces heating with a power of 2.5 kW; or cold air for air conditioning, with a power of 4 kW from solar energy. This prototype has a volume of about 1 m^3 and stores 40 kWh of cold or 60 kWh of heat, respectively, in the air conditioning and heating modes. The experiment was conducted using $SrBr_2$ and ammonia.

The recent development of hybrid cycles using a turbine also allows for the production of mechanical work to be proposed in conjunction with a production of heat or cold.

4.5.4. *Integration with a microgrid*

The use of the hydrogen vector for the energy transition has been an emerging technology for several years. The use of an electrolyzer and/or a fuel cell effectively makes it possible to switch from the electricity vector to the hydrogen vector without emitting any pollutants. However, these two components have rather low yields on the order of 70% for an electrolyzer and 50% for a fuel cell. Thus, a significant amount of heat is emitted when these systems operate. This heat, at a low temperature most of the time (<100°C), can be reused and/or stored in a sorption system. Oro et al. (2018) present a study where the recovery of the heat from a fuel cell allows the yield for the fuel cell to be increased from 50% to 63%.

4.6. Conclusion

This chapter has examined the storage of heat by chemical sorption from several different angles. The operation of this type of system has been detailed through the description of the chemical sorption reaction, how it operates, its equilibrium, its kinetics, as well as a description of the most common reagents used in the scientific literature. Then, through providing a broader level of description, the different types of reactors were presented, as well as the different types of modeling that can be found in the literature. Finally, some of basic processes have been described in order to familiarize the reader with these systems. In sections 4.4 and 4.5, we discussed some of the improvements made in storage processes, particularly through to the use of thermodynamic cycles with increased performance. Finally, the most obvious applications of chemical sorption systems have been presented, but it is clear that new applications will be found in the future since sorption processes have established themselves to be a promising new technology for confronting the climate challenge that humanity will have to face.

4.7. References

Bao, H.S., Wang, R.Z., Oliveira, R.G., Li, T.X. (2012). Resorption system for cold storage and long-distance refrigeration. *Applied Energy*, 93, 479–487. doi: 10.1016/j.apenergy.2011.12.022.

Boulnois, G. (2019). Intégration d'un procédé de stockage thermochimique à une centrale solaire thermodynamique : de l'expérimentation à l'échelle matériau aux performances énergétiques à l'échelle système. *HAL Open Science* [Online]. Available at: https://www.theses.fr/2016PERP0013.

Critoph, R.E. (1994). Forced convection enhancement of adsorption cycles. *Heat Recovery Systems and CHP*, 14(4), 343–350.

Critoph, R.E. (2012). Solid sorption cycles: A short history. *International Journal of Refrigeration*, 35(3), 490–493. doi: 10.1016/j.ijrefrig.2012.02.007.

Dąbrowski, A. (2001). Adsorption – From theory to practice. *Advances in Colloid and Interface Science*, 93(1/3), 135–224. doi: 10.1016/S0001-8686(00)00082-8.

Fadhel, M.I., Sopian, K., Daud, W.R.W. (2010). Performance analysis of solar-assisted chemical heat-pump dryer. *Solar Energy*, 84(11), 1920–1928. doi: 10.1016/j.solener.2010.07.001.

Faraday, M. (1823). On the condensation of several gases into liquids. *Philosophical Transactions of the Royal Society of London*, 113(1823), 189–198 [Online]. Available at: https://www.jstor.org/stable/107648 [Accessed 9 April 2024].

Ferrucci, F., Stitou, D., Ortega, P., Lucas, F. (2018). Mechanical compressor-driven thermochemical storage for cooling applications in tropical insular regions. Concept and efficiency analysis. *Applied Energy*, 219, 240–255. doi: 10.1016/j.apenergy.2018.03.049.

Fitó, J., Coronas, A., Mauran, S., Mazet, N., Perier-Muzet, M., Stitou, D. (2019). Hybrid system combining mechanical compression and thermo-chemical storage of ammonia vapor for cold production. *Energy Conversion and Management*, 180, 709–723. doi: 10.1016/j.enconman.2018.11.019.

Godefroy, A., Perier-Muzet, M., Mazet, N. (2019). Thermodynamic analyses on hybrid sorption cycles for low-grade heat storage and cogeneration of power and refrigeration. *Applied Energy*, 255, 113751. doi: 10.1016/j.apenergy.2019.113751.

Lahmidi, H., Mauran, S., Goetz, V. (2006). Definition, test and simulation of a thermochemical storage process adapted to solar thermal systems. *Solar Energy*, 80(7), 883–893. doi: 10.1016/j.solener.2005.01.014.

Le Pierrès, N., Mazet, N., Stitou, D. (2007a). Experimental results of a solar powered cooling system at low temperature. *International Journal of Refrigeration*, 30(6), 1050–1058. doi: 10.1016/j.ijrefrig.2007.01.002.

Le Pierrès, N., Stitou, D., Mazet, N. (2007b). New deep-freezing process using renewable low-grade heat: From the conceptual design to experimental results. *Energy*, 32(4), 600–608. doi: 10.1016/j.energy.2006.07.024.

Li, T.X., Wang, R.Z., Kiplagat, J.K., Chen, H. (2010). Experimental study and comparison of thermochemical resorption refrigeration cycle and adsorption refrigeration cycle. *Chemical Engineering Science*, 65(14), 4222–4230. doi: 10.1016/j.ces.2010.04.022.

Li, T.X., Wang, R.Z., Kiplagat, J.K., Kang, Y.T. (2013). Performance analysis of an integrated energy storage and energy upgrade thermochemical solid-gas sorption system for seasonal storage of solar thermal energy. *Energy*, 50(1), 454–467. doi: 10.1016/j.energy.2012.11.043.

Lu, H.B., Mazet, N., Spinner, B. (1996). Modelling of gas-solid reaction – Coupling of heat and mass transfer with chemical reaction. *Chemical Engineering Science*, 51(15), 3829–3845. doi: 10.1016/0009-2509(96)00010-3.

Lu, H.B., Mazet, N., Coudevylle, O., Mauran, S. (1997). Comparison of a general model with a simplified approach for the transformation of solid-gas media used in chemical heat transformers. *Chemical Engineering Science*, 52(2), 311–327. doi: 10.1016/S0009-2509(96)00407-1.

Lu, Z.S., Wang, R.Z., Li, T.X., Wang, L.X., Chen, C.J. (2007). Experimental investigation of a novel multifunction heat pipe solid sorption icemaker for fishing boats using CaCl 2/activated carbon compound–ammonia. *Engineering, Environmental Science*, 30, 76–85. doi: 10.1016/j.ijrefrig.2006.07.001.

Lua, Y.J., Bao, H.S., Yuan, Y., Wang, Y.D., Wang, L.W., Roskilly, A.P. (2014). Optimisation of a novel resorption cogeneration using mass and heat recovery. *Energy Procedia*, 61, 1103–1106. doi: 10.1016/j.egypro.2014.11.1032.

Mauran, S., Lahmidi, H., Goetz, V. (2008). Solar heating and cooling by a thermochemical process. First experiments of a prototype storing 60 kWh by a solid/gas reaction. *Solar Energy*, 82(7), 623–636. doi: 10.1016/j.solener.2008.01.002.

Mazet, N., Amouroux, M., Spinner, B. (1991). Analysis and experimental study of the transformation of a non-isothermal solid/gas reacting medium. *Chemical Engineering Communications*, 99(1), 155–174. doi: 10.1080/00986449108911585.

Michel, B., Mazet, N., Mauran, S., Stitou, D., Xu, J. (2012). Thermochemical process for seasonal storage of solar energy: Characterization and modeling of a high density reactive bed. *Energy*, 47(1), 553–563. doi: 10.1016/j.energy.2012.09.029.

Oro, M.V., de Oliveira, R.G., Bazzo, E. (2018). An integrated solution for waste heat recovery from fuel cells applied to adsorption systems. *Applied Thermal Engineering*, 136, 747–754. doi: 10.1016/j.applthermaleng.2018.01.081.

Perrigot, A., Perier-Muzet, M., Ortega, P., Stitou, D. (2020). Technical economic analysis of PV-driven electricity and cold cogeneration systems using particle swarm optimization algorithm. *Energy*, 211, 119009. doi: 10.1016/j.energy.2020.119009.

Pons, M. (1997). Global analysis of refrigerative adsorption cycles with thermal regeneration (non-uniform temperature). *International Journal of Refrigeration*, 20(6), 411–420. doi: 10.1016/S0140-7007(97)00035-2.

Ritter, J. (2008). Adsorption processes – History and recent developments. In *2008 AIChE Annual Meeting. Plenary III: 100 Years of Separations Science, Technology and Education*, 17–21 November, Philadelphia, PA.

Stitou, D., Goetz, V., Spinner, B. (1997). A new analytical model for solid-gas thermo-chemical reactors based on thermophysical properties of the reactive medium. *Chemical Engineering and Processing: Process Intensification*, 36(1), 29–43. doi: 10.1016/S0255-2701(96)04173-6.

Stitou, D., Mazet, N., Bonnissel, M. (2003). Performance of a high temperature hydrate solid/gas sorption heat pump used as topping cycle for cascaded sorption chillers. *Energy*, 29(2), 267–285. doi: 10.1016/j.energy.2003.08.011.

Stitou, D., Mazet, N., Mauran, S. (2012). Experimental investigation of a solid/gas thermochemical storage process for solar air-conditioning. *Energy*, 41(1), 261–270. doi: 10.1016/j.energy.2011.07.029.

Wang, L.W., Wang, R.Z., Oliveira, R.G. (2009). A review on adsorption working pairs for refrigeration. *Renewable and Sustainable Energy Reviews*, 13(3), 518–534. doi: 10.1016/J.RSER.2007.12.002.

Willers, E. and Groll, M. (1999). Evaluation of metal hydride machines for heat pumping and cooling applications. *International Journal of Refrigeration*, 22(1), 47–58. doi: 10.1016/S0140-7007(97)00056-X.

Yang, Z., Qu, M., Gluesenkamp, K.R. (2020). Design screening and analysis of gas-fired ammonia-based chemisorption heat pumps for space heating in cold climate. *Energy*, 207, 118213. doi: 10.1016/j.energy.2020.118213.

List of Authors

Kevyn JOHANNES
CETHIL
CNRS
Université Claude Bernard Lyon 1
Villeurbanne
France

Frédéric KUZNIK
CETHIL
CNRS
INSA Lyon
Villeurbanne
France

Nolwenn LE PIERRÈS
LOCIE
CNRS
Université Savoie Mont Blanc
Le Bourget-du-Lac
France

Lingai LUO
LTEN
CNRS
Nantes
France

Philippe MARTY
LEGI
CNRS
Grenoble INP
Université Grenoble Alpes
France

Kokouvi Edem N'TSOUKPOE
2iE
Ouagadougou
Burkina Faso

Maxime PERIER-MUZET
PROMES
CNRS
UPVD
Perpignan
France

Antoine PERRIGOT
PROMES
CNRS
UPVD
Perpignan
France

Larysa RATEL
CETHIL
CNRS
INSA Lyon
Villeurbanne
France

Driss STITOU
PROMES
CNRS
UPVD
Perpignan
France

Index

D

E

N

P

R

V, W

Z

Summary of Volume 1

Chapter 6. Latent Heat Storage Systems: Concepts and Applications
Marie DUQUESNE and Wahbi JOMAA

Chapter 7. Use of Hydrates for Cold Storage and Distribution in Refrigeration and Air-Conditioning Applications
Anthony DELAHAYE and Laurence FOURNAISON

Chapter 8. Concentrated Solar Power Plants and Storage
Régis OLIVÈS

Printed and bound by CPI Group (UK) Ltd, Croydon, CR0 4YY

13/09/2026

14961001-0001